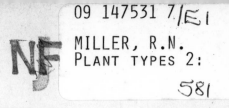

D1440034

70p

GREENWICH LIBRARIES
BOOK SALE

Mosses, ferns, conifers and flowering plants

Also from Hutchinson
Animal Types 1: Invertebrates
Animal Types 2: Vertebrates
M.A. Robinson and J.F. Wiggins

Plant Types 2

Mosses, ferns, conifers and flowering plants

Ruth N. Miller

HUTCHINSON

London Melbourne Sydney Auckland Johannesburg

Hutchinson & Co. (Publishers) Ltd
An imprint of the Hutchinson Publishing Group
17-21 Conway Street, London W1P 6JD

Hutchinson Publishing Group (Australia) Pty Ltd
16-22 Church Street, Hawthorn, Melbourne, Victoria 3122

Hutchinson Group (NZ) Ltd
32-34 View Road, PO Box 40-086, Glenfield, Auckland 10

Hutchinson Group (SA) (Pty) Ltd
PO Box 337, Bergvlei 2012, South Africa

First published 1985
Reprinted 1985

Set in 9/10 Univers by Words & Pictures Limited, London SE19

Printed and bound in Great Britain by
Anchor Brendon Ltd, Tiptree, Essex

British Library Cataloguing in Publication Data
Miller, Ruth N.
  Plant types.
  2: Bryophytes, pteridophytes and
  spermatophytes.
  1. Botany
  I. Title
581 QK47

ISBN 0 09 147531 7

Contents

Division Bryophyta (Liverworts and mosses)   8

Class Hepaticae   10
  Order Marchantiales   12
         Genus *Marchantia*   13
  Order Metzgeriales   15
         Genus *Pellia*   16
  Order Jungermanniales   18
         Genus *Lophocolea*   19

Class Musci   21
  Order Sphagnales   Genus *Sphagnum*   24
  Order Polytrichales   Genus *Polytrichum*   26
  Order Funariales   Genus *Funaria*   27
  Order Eubryales   Genus *Mnium*   30

Division Pteridophyta (Clubmosses, ferns and horsetails)   31

Class Lycopsida   33
  Order Lycopodiales   33
         Genus *Lycopodium*   35
  Order Selaginellales   37
         Genus *Selaginella*   39

Class Sphenopsida   41
  Order Equisetales   41
         Genus *Equisetum*   42

Class Pteropsida   44
  Order Filicales   44
         Genus *Hymenophyllum*   46
         Genus *Dryopteris*   47
         Genus *Pteridium*   49
  Some other Filicales   50

Division Spermatophyta (Seed-bearing plants)   51

Class Gymnospermae   53
  Order Cycadales   54
         Genus *Cycas*   56
  Order Coniferales   58
         Genus *Pinus*   60

Class Angiospermae   65
General structure of the Angiospermae   69
Branching   70
Leaf forms   71
Inflorescences   72
         Racemose inflorescences   74
         Cymose and mixed
         inflorescences   75

Contents

Floral morphology 76
Development of the androecium 79
Development of the ovule 80
The form of the gynaecium 82
Pollination 85
Fertilization and seed formation 87
Fruits 90
Dispersal 92
False fruits 95

Sub-class Monocotyledones 96
  Family Iridaceae 96
    Genus *Crocus* 97
  Family Liliaceae 99
    Genus *Endymion* 100
  Family Amaryllidaceae 101
    Genus *Narcissus* 102
  Family Graminae 104
    Genus *Poa* 105

Sub-class Dicotyledones 106
  Family Ranunculaceae 106
    Genus *Ranunculus* 107
  Family Cruciferae 108
    Genus *Cheiranthus* 109
  Family Leguminosae 110
    Genus *Cytisus* 111
  Family Rosaceae 112
    Genus *Malus* 113
  Family Scrophulariaceae 114
    Genus *Digitalis* 115
  Family Labiatae 116
    Genus *Lamium* 117
  Family Compositae 118
    Genus *Taraxacum* 119

The geological time scale 120

Glossary 123

Plant Types 2 covers the structure and life cycles of the liverworts and mosses, ferns and seed-bearing plants. It follows the same format as *Plant Types 1* in the way in which the main characteristics of each division are described and specific types chosen as examples. The types used as examples have been chosen to be representative of the range of forms within their division and relevant to the syllabuses of most GCE Examination Boards. In addition, *Plant Types 2* treats the floral and vegetative morphology of the angiosperms in some detail.

Division Bryophyta
Liverworts and mosses

Mainly terrestrial; a few aquatic.

Although considered to be land plants, the Bryophyta possess little in the way of cuticle and consequently are unable to withstand much desiccation; most genera are restricted in their distribution to damp situations. Some genera, such as *Tortula* found growing on walls, can survive in dry conditions, but no growth will take place unless moisture is present. A few species, such as *Fontinalis antipyretica* are entirely aquatic.

The persisting plant body is the gametophyte, a thalloid structure, showing no organization into stem, root and leaf comparable with that found in higher plants.

The form of the gametophyte can range from a simple, flat thallus in some of the liverworts, to a main stem-like axis-bearing leaf-like appendage in the mosses. All types bear rhizoids which serve to attach the plant to the substratum and also absorb water and nutrients. There is little internal cell differentiation in any members of this division.

Autotrophic nutrition; photosynthetic pigments present.

The Bryophyta possess chlorophylls a and b, beta-carotene and xanthophyll, identical with the photosynthetic pigments found in higher plants. Starch is formed as a food reserve.

Heteromorphic alternation of generations in the life cycle.

The haploid gametophyte generation is dominant and alternates with a diploid sporophyte generation which is attached to the gametophyte and is either wholly or partially dependent on it for its nutrition. Sexual reproduction is oogamous. Multicellular sex organs, male antheridia and female archegonia, develop on the gametophyte. The male gametes, the antherozoids, are biflagellate and fertilization is zoidogamous. The diploid zygote develops into the sporophyte generation, or sporogonium, and consists of an absorptive foot, a stalk or seta, and a capsule in which meiosis occurs prior to the formation of haploid spores. The Bryophyta are homosporous, only one type of spore being produced. In some cases, spore dispersal is aided by spirally thickened hygroscopic structures called elaters present in the capsule. In the liverworts, germination of the spores gives rise directly to the gametophyte, but in the mosses a characteristic filamentous stage, the protonema, is formed on which the gametophytes develop from buds.

Vegetative propagation by means of gemmae or regeneration of fragments of the thallus.

Gemmae, found in several mosses and many liverworts, are undifferentiated, multicellular structures, which develop on the gametophyte, often in cup-shaped receptacles. On dispersal from the parent plant, the gemmae germinate to give new plants. Regeneration from fragments of the gametophyte thallus is not uncommon in many members of this division.

The Bryophyta are considered to be primitive land plants. The scanty fossil record suggests that the group evolved in the Palaeozoic era, about 350 million years ago. They probably evolved from algal forms, similar to the Chlorophyta. It is unlikely that they ever formed a very large part of the earth's vegetation.

The classification is based on the nature of the thallus, the presence or absence of a protonema and the method of opening of the capsule.
There are two classes in this division:
Class Hepaticae – the liverworts
Class Musci – the mosses

Class Hepaticae
Liverworts

Most members of this class are terrestrial and there are only a few aquatic species. Liverworts found in temperate regions are restricted to shady, damp situations such as bogs, cliffs, woods and banks. Many species are found at the bases of tree trunks and on rotting logs. Species of *Riccia* are aquatic and grow on the surface of ponds rich in mineral nutrients. A large number of species are tropical. They thrive in the high humidity and are frequently found as epiphytes on forest trees.

The gametophyte thallus is dorsiventral, usually prostrate, and either thalloid or leafy. Amongst those liverworts that have a thalloid gametophyte, some show little cellular differentiation, as in *Pellia epiphylla*, whereas in members of the Marchantiales there is much greater cell specialization with definite photosynthetic tissue and air chambers. A large number of the liverworts show organization of the thallus into an axis on which leaf-like appendages are borne. The 'leaves' may be simple or lobed plates of cells and are usually arranged in three rows. The 'leaves' in the two rows on the dorsal surface are fully-developed, but those in the row on the ventral surface remain small. Dichotomous branching of the thallus is common and the rhizoids, which arise on the undersurface, are unicellular.

There is a heteromorphic alternation of generations in which the sporophyte is totally dependent on the gametophyte. Antheridia and archegonia develop on the gametophyte. Some species are dioecious, with antheridia and archegonia formed on different plants, while others are monoecious, with both male and female organs borne on the same plant. In the Marchantiales, the archegonia and antheridia are borne on stalks, or gametangiophores, which carry them above the surface of the thallus. In other orders, the antheridia either develop superficially on the upper surface of the thallus, or in the axils of the 'leaves'. The archegonia are lateral, or develop at the apex of the main 'shoot' in the leafy forms. Typically, each antheridium arises from a single cell and consists of a central mass of antherozoid mother cells surrounded by a single-layered protective wall and supported on a short stalk. The antherozoid mother cells divide mitotically to give rise to two naked, biflagellate antherozoids. Archegonia are formed from superficial initial cells and consist of a neck, made up of a number of tiers of cells, leading to a basal portion, the venter, which surrounds the oosphere. During development, the tip of the neck is closed and the central space is occupied by neck canal cells, which disintegrate at maturity, leaving a mucilage-filled canal down which the antherozoids swim prior to fertilization.

After fertilization, the diploid zygote develops into the sporophyte generation, often referred to as the sporogonium. It is a simple structure, consisting of a foot, which is embedded in gametophyte tissue and absorbs nutrients from the gametophyte, a seta or stalk and a capsule within which meiosis occurs prior to the development of the haploid spores.

During development, the sporogonium remains inside the basal part of the archegonium, which enlarges to accommodate it and is now called the calyptra. The mature capsule has very little sterile tissue, consisting mainly of haploid spores and elaters. On rupture, or dehiscence, of the capsule, the elaters, with their spiral thickenings, shrink irregularly as they dry and their contractions help to loosen and disperse the spore mass. The spores germinate in damp conditions to give short filaments of cells which develop into gametophytes. There is no protonemal stage.

Vegetative propagation is by means of gemmae or fragments of the gametophyte thallus.

Of the six orders in this class, representatives from three of the most common have been chosen:

Order Marchantiales   *Marchantia polymorpha*
Order Metzgeriales   *Pellia epiphylla*
Order Jungermanniales *Lophocolea* spp.

Order Marchantiales

This order includes both terrestrial and aquatic representatives.

The gametophytes of terrestrial genera are found growing over the substratum. They are commonly dorsiventrally flattened, ribbon-shaped and show dichotomous branching. Some species appear to have a simple external structure but show a very complex structure internally. In such species, the dorsal portion of the thallus is differentiated into air chambers which open to the atmosphere by means of simple pores. Within the air chambers is the main photosynthetic tissue. The ventral portion of the gametophyte thallus is made up of storage parenchyma with a simple conducting tissue of elongated cells. In most genera, there are transverse scales on the ventral surface and two types of rhizoids, some smooth-walled and others having peg-like invaginations.

The reproductive structures of the gametophyte generation vary in complexity within the order. The most elaborate gametangiophores are found in *Marchantia polymorpha*, where both antheridiophores and archegoniophores have a characteristic umbrella-like appearance. The archegonia are situated in rows on the undersurface of the archegoniophore, each row being separated from its neighbour by an outgrowth of tissue called the perichaetium. The antheridia are found in pits on the upper surface of the antheridiophore. *Marchantia polymorpha* is dioecious and fertilization is only possible if male and female plants are grown together.

The sporogonia are dependent on the gametophyte generation for their nutrition, and, in most members of the order, show the characteristic division into foot, seta and capsule. Within the capsule, meiosis takes place in the spore mother cells giving rise to tetrads of haploid spores.

Vegetative propagation in this order can either occur by fragmentation of the gametophyte thallus or by the formation of gemmae in some genera. Gemmae are flat, multicellular plates of cells, which develop on short stalks inside cup-like structures on the upper surface of the thallus. When mature, the gemmae become detached and are easily dispersed, germinating to form a new thallus in suitable conditions.

Marchantia polymorpha

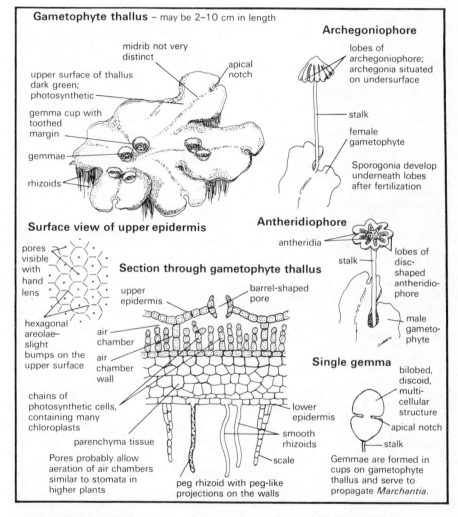

Gametophyte thallus – may be 2–10 cm in length

midrib not very distinct

apical notch

upper surface of thallus dark green; photosynthetic

gemma cup with toothed margin

gemmae

rhizoids

Archegoniophore

lobes of archegoniophore; archegonia situated on undersurface

stalk

female gametophyte

Sporogonia develop underneath lobes after fertilization

Surface view of upper epidermis

pores visible with hand lens

hexagonal areolae–slight bumps on the upper surface

chains of photosynthetic cells, containing many chloroplasts

parenchyma tissue

Pores probably allow aeration of air chambers similar to stomata in higher plants

Section through gametophyte thallus

upper epidermis

barrel-shaped pore

air chamber

air chamber wall

lower epidermis

smooth rhizoids

scale

peg rhizoid with peg-like projections on the walls

Antheridiophore

antheridia

stalk

lobes of disc-shaped antheridiophore

male gametophyte

Single gemma

bilobed, discoid, multi-cellular structure

apical notch

stalk

Gemmae are formed in cups on gametophyte thallus and serve to propagate *Marchantia*.

Marchantia polymorpha is a large, dark green, dichotomously branching liverwort, common on heaths, moors, banks and rocks in damp situations. It is also common in greenhouses and on cinder paths, and will grow particularly abundantly after there has been a fire.

The internal organization is complex. The main photosynthetic tissue is situated in air chambers on the upper surface of the thallus. These chambers have pores similar to the stomata of higher plants and it is thought that this arrangement allows gaseous exchange with a reduction in the amount of water lost by evaporation. Male and female plants are easily distinguished by their characteristic gametangiophores.

Marchantia polymorpha

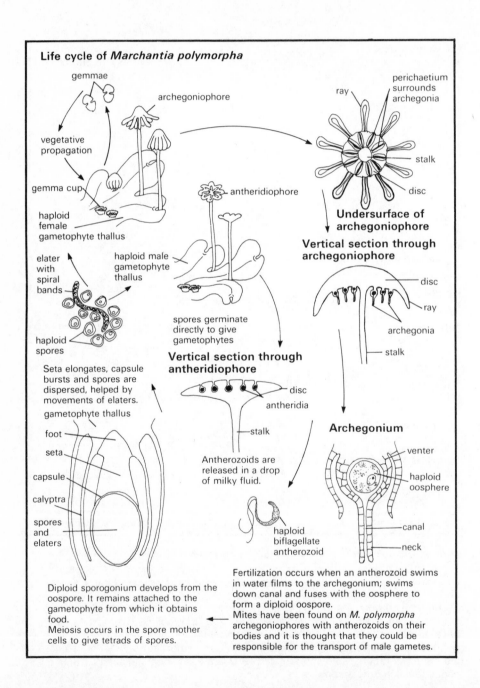

Life cycle of *Marchantia polymorpha*

gemmae

archegoniophore

vegetative propagation

gemma cup

haploid female gametophyte thallus

elater with spiral bands

haploid male gametophyte thallus

haploid spores

antheridiophore

spores germinate directly to give gametophytes

ray

perichaetium surrounds archegonia

stalk

disc

Undersurface of archegoniophore

Vertical section through archegoniophore

disc

ray

archegonia

stalk

Vertical section through antheridiophore

disc

antheridia

stalk

Antherozoids are released in a drop of milky fluid.

Seta elongates, capsule bursts and spores are dispersed, helped by movements of elaters.

gametophyte thallus

foot

seta

capsule

calyptra

spores and elaters

Archegonium

venter

haploid oosphere

canal

neck

haploid biflagellate antherozoid

Diploid sporogonium develops from the oospore. It remains attached to the gametophyte from which it obtains food.
Meiosis occurs in the spore mother cells to give tetrads of spores.

Fertilization occurs when an antherozoid swims in water films to the archegonium; swims down canal and fuses with the oosphere to form a diploid oospore.
Mites have been found on *M. polymorpha* archegoniophores with antherozoids on their bodies and it is thought that they could be responsible for the transport of male gametes.

Order Metzgeriales

The members of this order are mostly thalloid, as in *Pellia* spp., but leaf-like appendages are present in *Fossombronia* spp. and erect lamellae occur on the upper surface of the uncommon genus *Petalophyllum*.

The gametophyte thallus is simple and undifferentiated, often branching dichotomously in a manner similar to that shown by members of the Marchantiales. In many genera there is a well-defined midrib. Simple, uniform rhizoids arise from the midrib on the ventral surface. There is very little internal cell differentiation. Chloroplasts are abundant in the dorsal and ventral layers of the thallus.

The sporophyte has a foot, seta and capsule, characteristic of the class. The seta in many genera is capable of rapid and considerable elongation. The wall of the capsule is two or more layers thick, and the spore walls often show elaborate sculpturing, a feature which has been used in the identification of species.

Pellia epiphylla

DIVISION BRYOPHYTA

CLASS HEPATICAE

ORDER METZGERIALES

GENUS *Pellia*

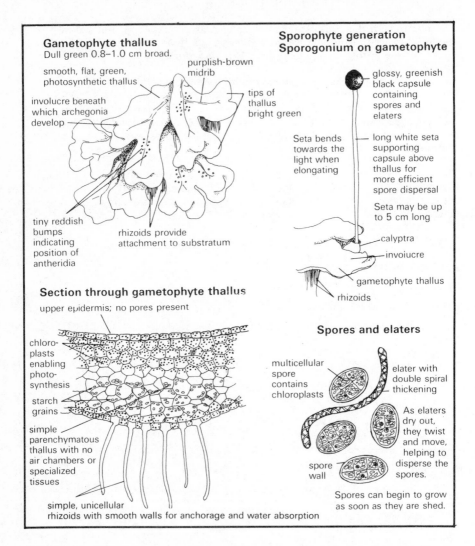

Gametophyte thallus
Dull green 0.8–1.0 cm broad.

smooth, flat, green, photosynthetic thallus

purplish-brown midrib

involucre beneath which archegonia develop

tips of thallus bright green

tiny reddish bumps indicating position of antheridia

rhizoids provide attachment to substratum

Sporophyte generation
Sporogonium on gametophyte

glossy, greenish black capsule containing spores and elaters

Seta bends towards the light when elongating

long white seta supporting capsule above thallus for more efficient spore dispersal

Seta may be up to 5 cm long

calyptra

involucre

gametophyte thallus

rhizoids

Section through gametophyte thallus

upper epidermis; no pores present

chloroplasts enabling photosynthesis

starch grains

simple parenchymatous thallus with no air chambers or specialized tissues

simple, unicellular rhizoids with smooth walls for anchorage and water absorption

Spores and elaters

multicellular spore contains chloroplasts

elater with double spiral thickening

As elaters dry out, they twist and move, helping to disperse the spores.

spore wall

Spores can begin to grow as soon as they are shed.

Pellia epiphylla is common on damp soil by streams and ditches. It grows in dense patches, dichotomously branched at the tips. It is a monoecious species, with both male and female sex organs produced on the same thallus. The sex organs begin to appear in early summer and, after fertilization, the sporogonium grows immediately. By autumn, it is well-developed, and then undergoes a period of dormancy for several months. In the spring of the following year, the seta will elongate, carrying the capsule above the surrounding vegetation. The capsule will dehisce under dry conditions, releasing the multicellular spores.

    P. epiphylla does not produce gemmae.

Pellia epiphylla

Life cycle of *Pellia epiphylla*

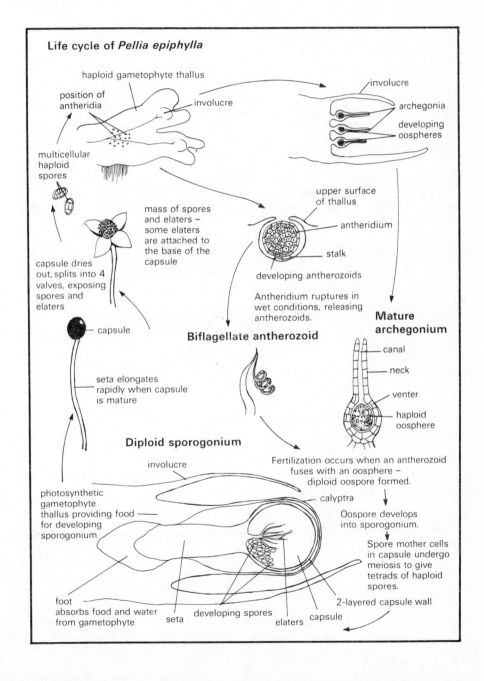

haploid gametophyte thallus

position of antheridia

involucre

involucre

archegonia

developing oospheres

multicellular haploid spores

mass of spores and elaters – some elaters are attached to the base of the capsule

upper surface of thallus

antheridium

stalk

developing antherozoids

Antheridium ruptures in wet conditions, releasing antherozoids.

Mature archegonium

capsule dries out, splits into 4 valves, exposing spores and elaters

capsule

Biflagellate antherozoid

canal

neck

venter

haploid oosphere

seta elongates rapidly when capsule is mature

Diploid sporogonium

involucre

photosynthetic gametophyte thallus providing food for developing sporogonium

calyptra

Fertilization occurs when an antherozoid fuses with an oosphere – diploid oospore formed.

Oospore develops into sporogonium.

Spore mother cells in capsule undergo meiosis to give tetrads of haploid spores.

foot absorbs food and water from gametophyte

seta

developing spores

elaters

capsule

2-layered capsule wall

17

Order Jungermanniales

The members of this order are mostly foliose, or leafy, but lack internal differentiation of tissues.

Growth of the gametophyte thallus is from an apical cell shaped like an inverted pyramid. Derivatives are cut off from this apical cell and thus the fundamental leaf arrangement is three-ranked. In many genera, the leaves forming a row along the mid-ventral line are smaller than those on the dorsal surface and are termed underleaves. It is quite usual for all the leaves to be lobed. Branching is never a true dichotomy, as in the Marchantiales and Metzgeriales, but varies in different genera and can be quite complex.

Each leaf-like appendage is a single layer of large isodiametric cells, containing chloroplasts and colourless oil bodies, and lacking a midrib. The stems are usually prostrate, delicate and show little internal differentiation. The rhizoids are simple and unicellular.

The archegonia develop directly from the apical cell, instead of behind it as in the other orders. They become surrounded by an erect, tube-like protective perianth, probably derived from the leaves. The antheridia arise in the axils of specialized leaves, or bracts, just behind the apex.

The sporophyte capsule may be spherical or ovoid, with a wall two or more layers thick. In this order, the spores tend to be small, spherical and show little sculpturing of the walls. The developing sporophyte is protected by the perianth of the gametophyte.

Lophocolea spp.

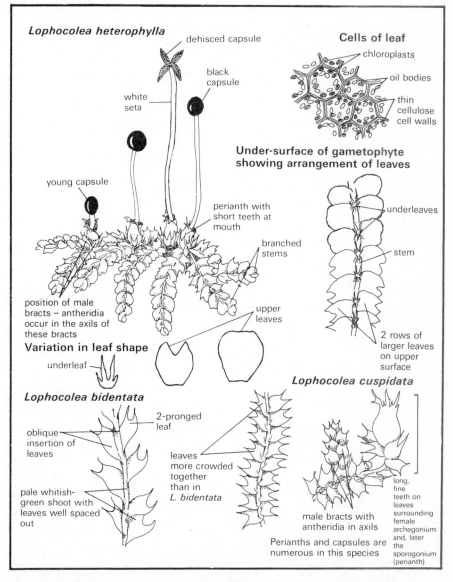

Lophocolea heterophylla

dehisced capsule

black capsule

white seta

young capsule

perianth with short teeth at mouth

branched stems

position of male bracts – antheridia occur in the axils of these bracts

upper leaves

Variation in leaf shape

underleaf

Lophocolea bidentata

oblique insertion of leaves

2-pronged leaf

pale whitish-green shoot with leaves well spaced out

leaves more crowded together than in *L. bidentata*

Cells of leaf

chloroplasts

oil bodies

thin cellulose cell walls

Under-surface of gametophyte showing arrangement of leaves

underleaves

stem

2 rows of larger leaves on upper surface

Lophocolea cuspidata

long, fine teeth on leaves surrounding female archegonium and, later the sporogonium (perianth)

male bracts with antheridia in axils

Perianths and capsules are numerous in this species

Lophocolea is a genus of leafy liverworts with prostrate, dorsiventral stems and characteristically bilobed leaves. The leaves are arranged in three ranks, those of the dorsal two ranks being large and fully developed, while those of the third ventral rank are small. *Lophocolea heterophylla*, found on decaying stumps and logs in moist woodlands, is probably the most conspicuous species. *L. bidentata* is found almost always on soil on town lawns or on the ground in woods, and *L. cuspidata* is common on rotting logs, branches and stumps as well as on the bases of living trees.

Class Musci
Mosses

Most members of this class are terrestrial, with only a few aquatic species. This is a very large group and has a wider distribution than the liverworts. The mosses may be the dominant vegetation in acid bogs, and in arctic and alpine regions. They are common in woodlands, hedgerows, some grasslands and even in urban areas on roofs, between paving stones and on walls.

The gametophyte thallus is upright or prostrate, showing organization into a stem-like axis with leaf-like appendages. The terms (leaf) and (stem) are used to describe the structures found in the moss gametophyte, but they are not homologous with the leaf and stem of higher plants. The bryophyte gametophyte is a haploid plant, whereas true stems and leaves are borne on diploid sporophytes in the higher plants.

The moss leaves, which are never lobed, usually consist of a single plate of cells spirally arranged around the stem. They often have a clearly-defined midrib, referred to as a nerve, which sometimes extends to form a hyaline apex to the leaf. Complex leaves, found in *Polytrichum* and related genera, have rows of parallel longitudinal lamellae, made up of cells containing chloroplasts, growing from the upper surface.

No true vascular tissue is found, but there is slightly more specialization of tissues in the mosses than there is in the liverworts. Some of the larger species have primitive conducting tissue for water and nutrients, the most complex arrangement being found in *Polytrichum*, where tracheid-like cells are present in the central part of the stem. Multicellular rhizoids with oblique cross walls are present and anchor the gametophyte to the substratum.

There is a heteromorphic alternation of generations. The sporophyte generation is usually only partially dependent on the gametophyte and often shows a high degree of differentiation. Both monoecious and dioecious species occur in this class and antheridia and archegonia develop apically in groups on the gametophyte, surrounded by leaves. In many genera, the leaves surrounding the groups of reproductive organs are not significantly different from the vegetative leaves, but in some cases, they are the largest leaves on the gametophyte. Those surrounding clusters of antheridia are known as perigonial leaves, and those surrounding the female archegonia are the perichaetial leaves. Their function, together with the sterile hairs or paraphyses which often occur, is protective and enables the retention of water films essential to the process of fertilization.

The sporophyte may possess chloroplasts and stomata, so that it can photosynthesise thus providing some of its own nutrients and making it less dependent on the gametophyte. The capsule is usually pear-shaped or cylindrical with a distinct lid; it contains more sterile tissue than the capsule of a liverwort. There are never any elaters produced in the mosses. In many species, when the capsule is mature, the lid is shed and a ring of tooth-like structures, the peristome, is exposed. The peristome teeth are hygroscopic and respond to changes in the humidity of the atmosphere, bending back when dry and curving inwards when wet. The movements of the teeth help in the dispersal of the spores. The spores germinate rapidly in damp conditions and give rise to protonemata, which are usually filamentous. Each protonema will develop rhizoids and buds, from which new leafy gametophytes grow.

Vegetative propagation is by means of gemmae or fragments of the gametophyte thallus. Gemmae are produced in a large number of genera. In some species, e.g. *Funaria hygrometrica*, green, filamentous growths, or secondary protonemata, can arise on stems, leaves or rhizoids. The tiny, bud-like structures which develop on the secondary protonemata give rise to new gametophytes. The term secondary protonemata is used for such filaments to distinguish them from the protonemata which arise on germination of the spores; these may be referred to as primary protonemata.

Some taxonomists recognize only three orders in this class: the Sphagnales, Andreaeales and Bryales. Others recognize these three groups as sub-classes, giving them the names Sphagnidae, Andreaeidae and Bryidae respectively. The Bryidae contains a large number of orders. The classification of mosses into these groups is based on differences in capsule structure and the nature of the protonema, rather than on gametophyte features which are very similar throughout the class.

The representatives illustrated here have been chosen from two sub-classes.
Sub-class Sphagnidae
> The ripe capsules are ovoid or globose in shape, and raised on a short stalk, which is part of the gametophyte. They usually lack a peristome and the spores are discharged explosively. The spores germinate to give small thalloid protonemata.
> Order Sphagnales   *Spagnum palustre*

Sub-class Bryidae
> The capsule is complex, consisting of a cylindrical region of spore-producing tissue around a central sterile columella. A peristome is present in most genera. The spores germinate to give filamentous protonematá.
> Order Polytrichales   *Polytrichum commune*
> Order Funariales   *Funaria hygrometrica*
> Order Eubryales   *Mnium hornum*

Summary of differences between Hepaticae and Musci

Hepaticae	Musci
Gametophyte thalloid or foliose (leafy)	Gametophyte foliose
Gametophyte usually prostrate in habit	Gametophyte prostrate or erect in habit
Dichotomous branching common	Dichotomous branching rare
Growth from an apical cell or row of cells	Growth always from an apical cell
Sex organs lateral or, more rarely, apical	Sex organs always apical
'Leaf', if present, has no nerve; usually lobed	'Leaf' often with nerve; never lobed
Rhizoids unicellular	Rhizoids multicellular with oblique cross walls
Sporophyte simple with little sterile tissue; no stomata or photosynthetic tissue; totally dependent on gametophyte	Sporophytes more complex with stomata and photosynthetic tissue; partially independent of the gametophyte
Capsule usually dehisces by splitting into valves; spore dispersal aided by elaters	Capsule protected by operculum which is forced off at maturity; spore dispersal aided by system of peristome teeth; elaters never present
Spores never develop into protonemata	Spores always develop into protonemata

Sphagnum spp.

DIVISION BRYOPHYTA

CLASS MUSCI

ORDER SPHAGNALES

GENUS *Sphagnum*

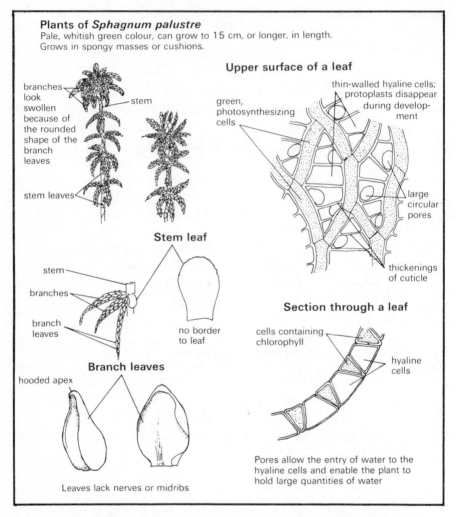

Plants of *Sphagnum palustre*
Pale, whitish green colour, can grow to 15 cm, or longer, in length. Grows in spongy masses or cushions.

branches look swollen because of the rounded shape of the branch leaves

stem

stem leaves

Upper surface of a leaf

green, photosynthesizing cells

thin-walled hyaline cells; protoplasts disappear during development

large circular pores

thickenings of cuticle

Stem leaf

stem

branches

branch leaves

no border to leaf

Branch leaves

hooded apex

Leaves lack nerves or midribs

Section through a leaf

cells containing chlorophyll

hyaline cells

Pores allow the entry of water to the hyaline cells and enable the plant to hold large quantities of water

The order Sphagnales is represented by the single genus *Sphagnum*, readily recognized by its branching system, its distinctive pattern of photosynthesizing and hyaline cells and its capacity to hold large quantities of water. All species grow in moist places, such as bogs, ponds and wet woodlands, particularly in acid conditions. The genus is important ecologically in the plant succession in acid, waterlogged habitats; the growth of the plants can gradually change an area of wet bog into drier peaty ground, which may eventually support woodland. As the plants grow upwards, the basal parts die, but do not decay because of the acid conditions. Masses of dead tissues become compacted together and, with other plant remains, eventually form peat.

Peat has been, and still is, used extensively as fuel. It is dug out in blocks and needs to be dried for several months before it can be burnt. Nursery-workers use peat for packing live plants, because it retains moisture, and it can be added to dry soils to improve their water-holding capacity. Because dry *Sphagnum* can absorb large quantities of moisture, it has been used as an absorbent in surgical dressings.

Polytrichum commune

DIVISION BRYOPHYTA

CLASS MUSCI

ORDER POLYTRICHALES

GENUS *Polytrichum*

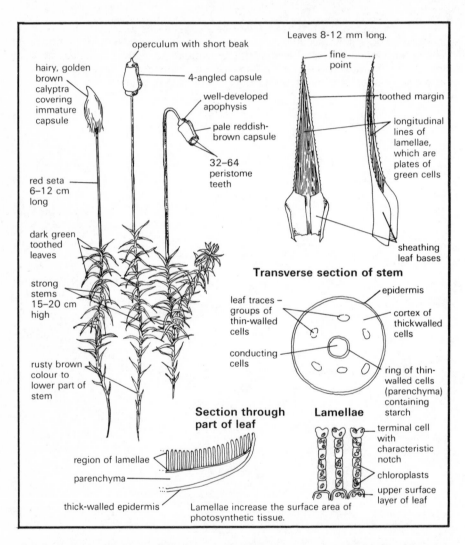

Leaves 8-12 mm long.

operculum with short beak

hairy, golden brown calyptra covering immature capsule

4-angled capsule

well-developed apophysis

pale reddish-brown capsule

32–64 peristome teeth

red seta 6–12 cm long

dark green toothed leaves

strong stems 15–20 cm high

rusty brown colour to lower part of stem

fine point

toothed margin

longitudinal lines of lamellae, which are plates of green cells

sheathing leaf bases

Transverse section of stem

epidermis

leaf traces – groups of thin-walled cells

conducting cells

cortex of thickwalled cells

ring of thin-walled cells (parenchyma) containing starch

Section through part of leaf

Lamellae

region of lamellae

parenchyma

thick-walled epidermis

terminal cell with characteristic notch

chloroplasts

upper surface layer of leaf

Lamellae increase the surface area of photosynthetic tissue.

Polytrichum species have little-branched, wiry, erect stems, which arise from a complex underground rhizoid system. The plants usually grow in large patches or tufts. They are found on heathland, in woodlands and some species are characteristic of acid bogs amongst the *Sphagnum*.

All the European species are dioecious and the male gametophytes can be distinguished by the pinkish rosette of leaves at the apex, which resembles a flower. All members of the genus have a very hairy calyptra covering the young capsule, from which the name *Polytrichum* is derived.

DIVISION BRYOPHYTA
CLASS MUSCI
ORDER FUNARIALES
GENUS *Funaria*

Funaria hygrometrica

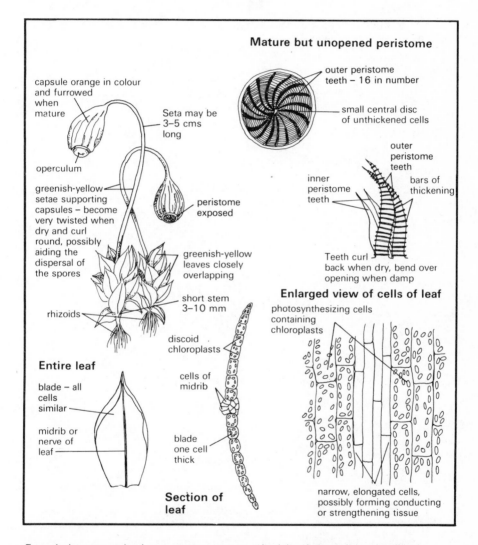

Mature but unopened peristome

outer peristome teeth – 16 in number

small central disc of unthickened cells

capsule orange in colour and furrowed when mature

Seta may be 3–5 cms long

operculum

greenish-yellow setae supporting capsules – become very twisted when dry and curl round, possibly aiding the dispersal of the spores

peristome exposed

greenish-yellow leaves closely overlapping

outer peristome teeth

inner peristome teeth

bars of thickening

Teeth curl back when dry, bend over opening when damp

rhizoids

short stem 3–10 mm

Enlarged view of cells of leaf

photosynthesizing cells containing chloroplasts

discoid chloroplasts

Entire leaf

blade – all cells similar

midrib or nerve of leaf

cells of midrib

blade one cell thick

Section of leaf

narrow, elongated cells, possibly forming conducting or strengthening tissue

Funaria hygrometrica is a common moss, colonizing bare soil in woodland, moorland and in gardens. It can form extensive carpets, particularly on soil that has recently been burnt. The appearance of *Funaria* spp. on bonfire sites has been linked with the high pH and nutrient content of an area after a fire.

There is very little internal tissue specialization in the stems and leaves of *Funaria* spp., in contrast to that found in *Polytrichum* spp. Although water and mineral salts are absorbed by the rhizoids, there seems to be little internal transport of water and water uptake can take place over the external surface of the whole plant.

No gemmae are formed, but secondary protonemata may develop on the plants as explained on page 22.

Funaria hygrometrica

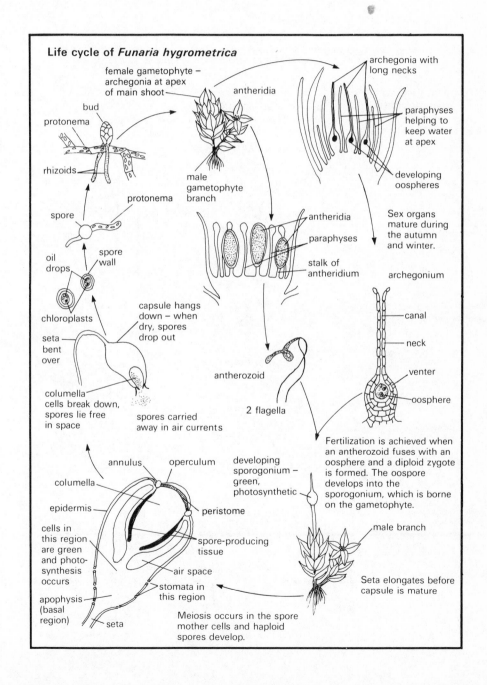

Life cycle of *Funaria hygrometrica*

female gametophyte – archegonia at apex of main shoot

antheridia

archegonia with long necks

paraphyses helping to keep water at apex

bud

protonema

rhizoids

developing oospheres

male gametophyte branch

spore

protonema

oil drops

spore wall

chloroplasts

antheridia

paraphyses

stalk of antheridium

Sex organs mature during the autumn and winter.

archegonium

canal

neck

venter

oosphere

seta bent over

capsule hangs down – when dry, spores drop out

columella cells break down, spores lie free in space

spores carried away in air currents

antherozoid

2 flagella

Fertilization is achieved when an antherozoid fuses with an oosphere and a diploid zygote is formed. The oospore develops into the sporogonium, which is borne on the gametophyte.

annulus

operculum

columella

epidermis

peristome

cells in this region are green and photo-synthesis occurs

apophysis (basal region)

seta

developing sporogonium – green, photosynthetic

spore-producing tissue

air space

stomata in this region

male branch

Seta elongates before capsule is mature

Meiosis occurs in the spore mother cells and haploid spores develop.

29

Mnium hornum

DIVISION BRYOPHYTA

CLASS MUSCI

ORDER EUBRYALES

GENUS Mnium

Mnium hornum is an abundant species, found growing in dark green tufts on the ground in woodland and on banks. It is characteristic of beechwoods and often occurs on grassy ledges of sea cliffs.

Mature sporophyte on gametophyte

pendulous capsule 5 mm long

pale yellow capsule with narrow red rim

yellowish peristome

reddish colour to seta 2.5–5 cm long

bright green leaves which curl up when dry

stem 2–4 cm long

rhizoids

Female gametophyte

leaves round archegonia

Apex of male gametophyte

rosette of leaves round antheridia

Section through capsule

seta

apophysis

air space

columella

spore-producing tissue

annulus

peristome

operculum

Double peristome

inner peristome with broad cilia and narrow cilia

outer peristome

Entire leaf
may be up to 4 mm long

nerve ends below extreme tip of leaf

elongated leaf

border

midrib made up of elongated cells with thick walls

'teeth'

lamina

one-celled projections form 'tooth' on leaf margin

leaf margin or border of long narrow cells with a reddish tinge

Division Pteridophyta

Mostly terrestrial; some aquatic and some epiphytic.

This group includes the ferns, the club-mosses and the horsetails, and it would seem that, with their larger and more complex sporophytes, the Pteridophyta are much better adapted to a terrestrial existence than the Bryophyta. However, in those species where the gametophytes are thalloid and liable to desiccation because they lack cuticle, the sporophytes are limited to damper habitats. In some species, these limitations do not exist, as the gametophytes are either subterranean or retained within the spore, and a much wider range of habitats can be exploited.

Persisting plant body is the sporophyte, which shows differentiation into true stem, leaf and root systems; gametophyte reduced.

Dichotomous branching in the leafy, perennial sporophytes is common, and there is a much higher degree of cell specialization than is found in any of the Bryophyta. Conducting tissues are present and consist of xylem, made up of tracheids, and phloem. The gametophyte is usually relatively short-lived and small by comparison with the sporophyte, and is simple in structure with little cell specialization. Some gametophytes are thalloid, resembling simple liverworts, while others are branched and filamentous, or tuberous, colourless and subterranean, feeding saprophytically. In the most advanced forms, the gametophytes are reduced to a few cells which remain enclosed within the spore wall.

Autotrophic nutrition; photosynthetic pigments are present.

The Pteridophyta possess photosynthetic pigments identical with those of higher plants. Starch, protein and lipids are stored as food reserves.

Heteromorphic alternation of generations in the life cycle.

The diploid sporophyte generation is dominant and alternates with the haploid gametophyte generation, from which it is totally independent when mature. The sporophyte bears sporangia which produce haploid spores. These spores may be all identical (homosporous) or of two kinds, microspores and megaspores (heterosporous). In homosporous species, the gametophytes, which develop on germination of the spores, are always nutritionally independent of the sporophytes, but this is not always the case in heterosporous species. In homosporous species, antheridia and archegonia are borne on the same gametophyte, or prothallus, but in the heterosporous species, the smaller microspores give rise to male prothalli bearing antheridia, and the larger megaspores give rise to female prothalli bearing archegonia. The antheridia and archegonia bear a close resemblance to those of the Bryophyta, arising from superficial cells near the growing points of the gametophytes. Each archegonium consists of a neck and a venter which surrounds the oosphere. The neck is made up of fewer tiers of cells than is usual in the Bryophyta. The antherozoids which are liberated from the antheridia are motile, biflagellate in the Lycopsida and multiflagellate in the Sphenopsida and Pteropsida. Mature antherozoids are released in moist conditions and swim in water films to the archegonia. Fertilization is achieved when an antherozoid fuses with the oosphere in an archegonium, and the process is still dependent on water (zoidogamous). After fertilization, the sporophyte is dependent on the gametophyte during its early stages of development, but becomes independent on the production of green, photosynthesizing leaves.

The Pteridophyta is an ancient group of plants with many fossil forms known, and it is thought from the evidence that they first appeared in the Palaeozoic era. Tree ferns and lycopods flourished in the Upper Devonian and Carboniferous periods, when it is thought that they formed large forests in swampy regions. They were the dominant vegetation 260 million years ago, and were replaced by the seed-bearing plants, the Spermatophyta.
There are five classes in this division:

Class Psilophytopsida all fossil forms   } not dealt with here
Class Psilotopsida   simplest Pteridophyta }
Class Lycopsida   club-mosses
Class Sphenopsida   horsetails
Class Pteropsida   applied here to the large-leaved ferns

Comparison of the three classes

Lycopsida	Sphenopsida	Pteropsida
Leaves small (microphyllous)	Leaves small (microphyllous)	Leaves large (macrophyllous)
Spiral leaf arrangement	Whorled leaf arrangement	Spiral leaf arrangement
Homosporous or heterosporous	Homosporous	Homosporous
Sporangia aggregated into cones or strobili	Sporangia borne on distinctive sporangiophores in a strobilus	Sporangia borne on underside or margins of leaves
No leaf gaps in the stele (vascular tissue)	No leaf gaps in the stele	Leaf gaps in the stele in most cases
Order Lycopodiales Homosporous, leaves without ligules (scales), e.g. *Lycopodium* spp.	**Order Equisetales** e.g. *Equisetum* spp.	**Order Filicales** e.g. *Hymenophyllum* spp. *Dryopteris* spp. *Pteridium* spp.
Order Selaginellales Heterosporous, leaves with ligules, e.g. *Selaginella* spp.		

Class Lycopsida
Order Lycopodiales

The members of this order are terrestrial, with a world-wide distribution. The majority of the members of this order are found growing in tropical regions, but some species do occur in alpine and arctic regions. In the colder climates, the plants have short erect stems, or creeping stems with erect lateral branches and are chiefly found growing on peaty soils in mountainous areas. The tropical species are much larger, often growing epiphytically on the trunks of trees.

The sporophytes are microphyllous; the gametophytes are often subterranean and saprophytic. The stem of the sporophyte is surrounded by the spirally-arranged small leaves, or microphylls, which have a single vein and simple outline. Stomata occur on both surfaces of the leaf, but there is no clear distinction between palisade and spongy mesophyll. Ligules are tiny scales occurring at the bases of the leaves in some Lycopsida, but are not present in this order, thus distinguishing the Lycopodiales from the Selaginellales. The spore-bearing leaves, or sporophylls, may be the same size and shape as the sterile leaves, as in *Lycopodium selago*, or they may be different in shape and organized into definite strobili, or cones, as in *L. clavatum*. The roots show dichotomous branching and are mostly adventitious, arising from the base of the stem in erect species, or at intervals along the creeping stem in the prostrate species. In the stem, the vascular tissues, or stele, may consist of a stellate core of tracheids with phloem between the arms, or the xylem and phloem may be arranged in alternate bands. In some cases, phloem and xylem may be completely mixed up with sieve cells occurring amongst the tracheids.

In subterranean gametophytes, or prothalli, nutrients are obtained from a mycorrhizal fungus, with which each prothallus must become associated at an early stage. If this does not happen, the prothalli die. Sometimes prothalli are partly exposed to the air, in which case the upper parts turn green.

There is a heteromorphic alternation of generations in the life cycle. Sporangia are borne on the upper surface or in the axils of the sporophylls, and consist of a central mass of spore mother cells surrounded by a wall several cells thick. The wall contains an inner nutritive layer, the tapetum, which breaks down to provide nutrients for the developing spores. Tetrads of haploid spores are produced after meiosis of the spore mother cells. Each spore develops a thick, cutinized wall and has a conspicuous tri-radiate ridge at its apex. The sporangia dehisce in the region of the thin-walled cells which make up the stomium, and the spores are dispersed by air movements. Germination is usually delayed for several years and when it does occur, the gametophyte may take several years to mature. Male and female sex organs are produced on the upper surface of the inverted cone-shaped prothallus. Biflagellate antherozoids develop in antheridia and are thought to be chemically attracted to the oospheres in the archegonia, where fertilization takes place. The resulting diploid zygote develops into an embryo, with an absorptive foot together with a root, first leaf and stem apex. The young sporophyte is nutritionally dependent on the gametophyte until it can photosynthesize and lead an independent existence.

Vegetative propagation by means of bulbils or gemmae may occur in some species. In *L. selago*, tiny leafy structures arise in the axils of leaves on young shoots. These bulbils or gemmae, on becoming detached from the parent plant, can develop roots and grow into new sporophytes.

Lycopodium spp.

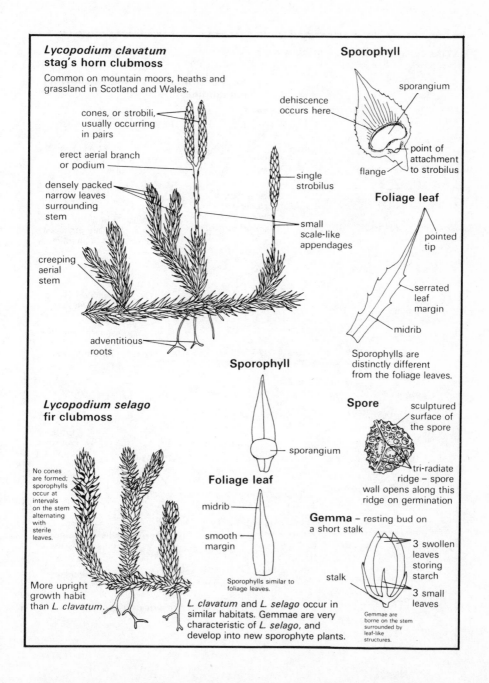

Lycopodium clavatum
stag's horn clubmoss

Common on mountain moors, heaths and grassland in Scotland and Wales.

cones, or strobili, usually occurring in pairs

erect aerial branch or podium

densely packed narrow leaves surrounding stem

creeping aerial stem

single strobilus

small scale-like appendages

adventitious roots

Sporophyll

sporangium

dehiscence occurs here

point of attachment to strobilus

flange

Foliage leaf

pointed tip

serrated leaf margin

midrib

Sporophylls are distinctly different from the foliage leaves.

Lycopodium selago
fir clubmoss

No cones are formed; sporophylls occur at intervals on the stem alternating with sterile leaves.

More upright growth habit than *L. clavatum*.

Sporophyll

sporangium

Foliage leaf

midrib

smooth margin

Sporophylls similar to foliage leaves.

L. clavatum and *L. selago* occur in similar habitats. Gemmae are very characteristic of *L. selago*, and develop into new sporophyte plants.

Spore

sculptured surface of the spore

tri-radiate ridge – spore wall opens along this ridge on germination

Gemma – resting bud on a short stalk

3 swollen leaves storing starch

stalk

3 small leaves

Gemmae are borne on the stem surrounded by leaf-like structures.

35

Lycopodium clavatum

DIVISION PTERIDOPHYTA

CLASS LYCOPSIDA

ORDER LYCOPODIALES

GENUS *Lycopodium*

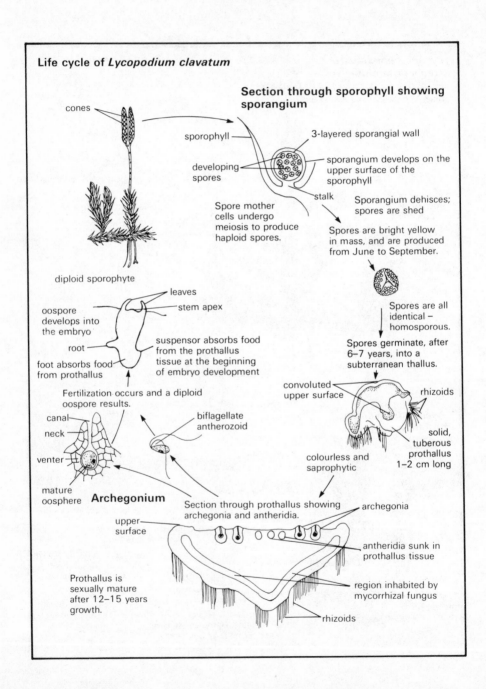

Life cycle of *Lycopodium clavatum*

cones

diploid sporophyte

Section through sporophyll showing sporangium

sporophyll

developing spores

3-layered sporangial wall

sporangium develops on the upper surface of the sporophyll

stalk

Spore mother cells undergo meiosis to produce haploid spores.

Sporangium dehisces; spores are shed

Spores are bright yellow in mass, and are produced from June to September.

Spores are all identical – homosporous.

Spores germinate, after 6–7 years, into a subterranean thallus.

convoluted upper surface

rhizoids

solid, tuberous prothallus 1–2 cm long

colourless and saprophytic

oospore develops into the embryo

leaves

stem apex

root

foot absorbs food from prothallus

suspensor absorbs food from the prothallus tissue at the beginning of embryo development

Fertilization occurs and a diploid oospore results.

canal

neck

venter

biflagellate antherozoid

mature oosphere

Archegonium

Section through prothallus showing archegonia and antheridia.

upper surface

archegonia

antheridia sunk in prothallus tissue

region inhabited by mycorrhizal fungus

Prothallus is sexually mature after 12–15 years growth.

rhizoids

Order Selaginellales

Members of this order are terrestrial, with a world-wide distribution. Many species in this order, which contains one single living genus, *Selaginella*, are found in the humid rainforests of the tropics, and range in size from small epiphytic forms to large climbing plants. Some species can tolerate very dry conditions and are found in desert regions, showing remarkable recovery after long periods of drought. There is only one native British species, *Selaginella selaginoides*, found in mountainous areas of Scotland and Wales, but *S. kraussiana*, a very common greenhouse plant, seems to have become naturalized in some areas of Britain.

The sporophytes are microphyllous. The gametophytes develop within the spore walls and are nutritionally dependent on the sporophytes. The stems of the sporophytes are more branched than those of the Lycopodiales, and the branches often grow out in the same plane, giving the plants a frond-like appearance. The microphylls, or leaves, are small, usually inserted spirally, but arranged in four rows. In *S. selaginoides*, they radiate around the stem and are of equal size, but in *S. kraussiana*, there are two rows of small leaves on the upper surface of the stem and two rows of larger leaves on the lower surface. Close to the base of the upper surface of each leaf is a tiny, scale-like flap of tissue called a ligule. The sporophylls are always aggregated into cones, or strobili. Special root-like structures, or rhizophores, occur in some species, e.g. *S. kraussiana*. These grow from the stem down towards the soil, forming a dichotomy when the soil surface is near, and developing roots once contact with the soil is made.

The internal anatomy is essentially similar to that found in the Lycopodiales, but *Selaginella* spp. have a unique type of endodermis, in which strands of hypha-like cells suspend the stele in a central cavity. This is called a trabeculate endodermis and does not appear to have any known physiological significance.

The female gametophyte consists of a group of cells formed beneath the tri-radiate ridge of the megaspore, and it is exposed on the splitting of the megaspore wall. There may be some development of chlorophyll and a few rhizoids are often produced, but the female prothallus never becomes detached from the spore. The male prothallus is even more reduced, consisting solely of a small prothallial cell cut off in the first division of the microspore. Neither prothallus is nutritionally independent.

There is a heteromorphic alternation of generations in the life cycle. The sporophylls are very similar in structure to the sterile microphylls, each having a ligule at the base. A stalked sporangium develops on the upper surface of the sporophyll between the ligule and the axis of the strobilus. *Selaginella* spp. are heterosporous and there are two kinds of sporangia: microsporangia producing microspores and megasporangia producing megaspores. In most species, both types of sporangia are found in the same strobilus, the microsporangia occurring near the apex, and the megasporangia near the base of the strobilus. The development of the microspores is similar to the development of spores in *Lycopodium* spp., but in the megasporangium, many of the megaspore mother cells abort so that few megaspores develop, the precise number varying with the species. The megaspores become very large with thick, sculptured walls, and they accumulate reserves of food. In both types of spore, development begins before they are shed from the sporophyte. The microspore undergoes division, resulting in the formation of a small prothallial cell and a large antheridial cell, which will continue to divide and form an antheridium containing 128 or 256 biflagellate antherozoids. In the megaspore, the female prothallus develops and bears archegonia in the central region below the triradiate ridge. Both types of spore rupture along this ridge. The microspores release their antherozoids, which swim in water to the archegonia of an exposed female prothallus. It is thought that the rhizoids on the female prothallus help to retain water in which the antherozoids congregate near the archegonia. Fertilization occurs and the resulting diploid zygote begins to develop into an embryo sporophyte. The sequence of events is similar to that found in *Lycopodium* spp., except that there is the development of a suspensor, a one- or few-celled structure, which pushes down into the female gametophyte tissue, bringing the embryo into contact with the food reserves. The thick megaspore wall prevents the spore from drying out and protects it from decay. The substantial food reserves of the megaspore enable considerable growth of the embryo to take place before it needs an external source of nutrients. There is a significant decrease in the length of the gametophyte phase in the life cycle of *Selaginella* spp. compared with that of *Lycopodium* spp.

Selaginella kraussiana

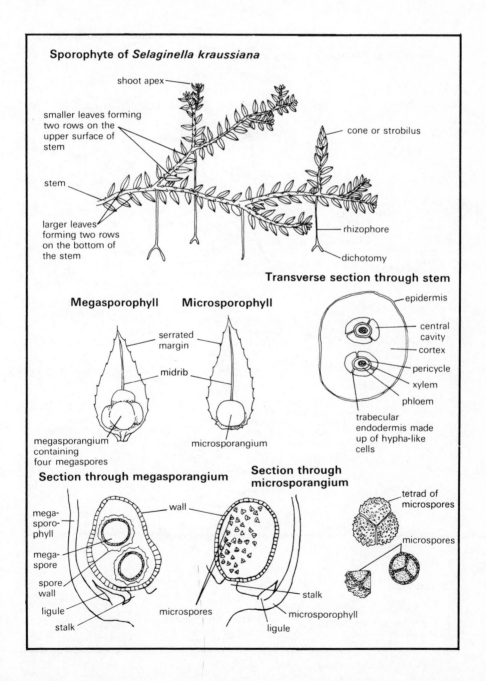

Sporophyte of *Selaginella kraussiana*

shoot apex

smaller leaves forming two rows on the upper surface of stem

cone or strobilus

stem

larger leaves forming two rows on the bottom of the stem

rhizophore

dichotomy

Transverse section through stem

Megasporophyll   **Microsporophyll**

serrated margin

midrib

megasporangium containing four megaspores

microsporangium

epidermis

central cavity

cortex

pericycle

xylem

phloem

trabecular endodermis made up of hypha-like cells

Section through megasporangium

Section through microsporangium

wall

mega-sporo-phyll

mega-spore

spore wall

ligule

stalk

microspores

tetrad of microspores

microspores

stalk

microsporophyll

ligule

39

Selaginella kraussiana

Life cycle of *Selaginella kraussiana*

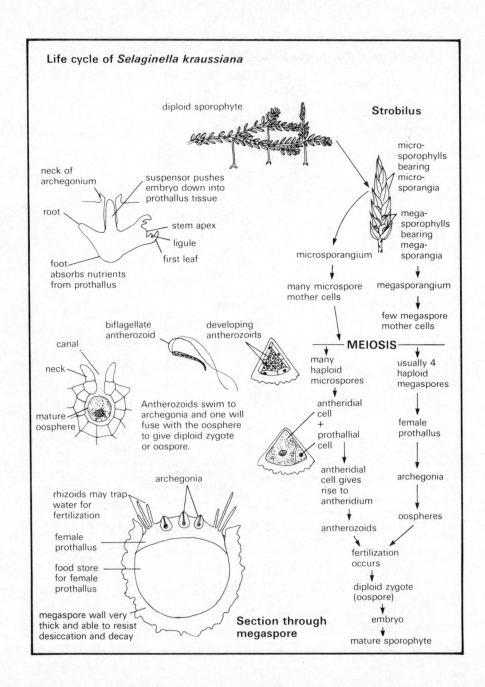

diploid sporophyte

Strobilus

micro-sporophylls bearing micro-sporangia

mega-sporophylls bearing mega-sporangia

neck of archegonium

suspensor pushes embryo down into prothallus tissue

root

stem apex

ligule

first leaf

foot absorbs nutrients from prothallus

microsporangium

megasporangium

many microspore mother cells

few megaspore mother cells

biflagellate antherozoid

developing antherozoids

canal

neck

MEIOSIS

many haploid microspores

usually 4 haploid megaspores

mature oosphere

Antherozoids swim to archegonia and one will fuse with the oosphere to give diploid zygote or oospore.

antheridial cell + prothallial cell

female prothallus

antheridial cell gives rise to antheridium

archegonia

archegonia

rhizoids may trap water for fertilization

female prothallus

food store for female prothallus

megaspore wall very thick and able to resist desiccation and decay

antherozoids

oospheres

fertilization occurs

diploid zygote (oospore)

Section through megaspore

embryo

mature sporophyte

Class *Sphenopsida*
Order *Equisetales*

The members of this order are terrestrial. The one living genus is *Equisetum*, and it is more commonly found in temperate zones than in the tropics. Some species grow in damp situations, along river banks and in marshy areas, but others can tolerate drier conditions such as are found on railway embankments and in urban gardens.

The sporophyte is microphyllous and the leaves are borne in whorls. The gametophyte is small, terrestrial and photosynthetic. All the species of this genus have a perennial underground rhizome, bearing two types of erect aerial shoots, which are annual in temperate and arctic regions. The vegetative stem contains photosynthetic tissue, has longitudinal furrows and bears whorls of brownish scale leaves, which are fused to form a sheath at each node. Lateral branches arise in whorls at each node. The fertile stem is similarly furrowed with scale leaves at the nodes, but it does not have whorls of branches, or chlorophyll, and has a single cone, or strobilus, at its apex. Each strobilus develops a large number of flat-stalked appendages, called sporangiophores, which bear 5 to 10 sporangia on their under-surface. The gametophyte prothalli, which develop from the spores, consist of branching flat plates of green tissue with numerous rhizoids growing from the lower surface. All the prothalli have a similar structure but some bear antheridia and others archegonia.

A mature sporophyte stem has a large central cavity which is surrounded by a ring of vascular bundles. The position of the bundles corresponds with the ribs on the stem, and there is usually a definite endodermis present. There are longitudinal canals, called vallecular canals, which are air-filled and positioned in the cortex between the endodermis and the epidermis. These canals alternate with the vascular bundles. The ribs of the stem consist of sclerenchyma tissue, the cells of which contain deposits of silica. This sclerenchyma contributes most to the strength of the stem, as there is not much xylem in the vascular bundles. The stomata are located in the furrows, and lie above the vallecular canals.

There is a heteromorphic alternation of generations in the life cycle. Sporangia develop on fertile shoots of the sporophyte, and the spores arise after meiosis of the spore mother cells. Each haploid spore secretes a wall, the outermost layer of which is made up of four spiral strips. These strips, or elaters, remain coiled around the spore until the sporangium dehisces, and then, on drying out, undergo quite violent movements assisting in the dispersal of the spores. At the dispersal stage, the spores are green due to the formation of chlorophyll. *Equisetum* spp. are homosporous, but two types of gametophyte prothalli develop; male prothalli bearing antheridia and female prothalli, which are larger than the male prothalli, bearing archegonia. The multiflagellate, spirally-coiled antherozoids swim in water to the oospheres in the archegonia and fertilization occurs. Unusually for this division, several sporophytes may develop on one prothallus after fertilization.

Vegetative propagation of the sporophyte occurs by means of tubers. In some species, the rhizomes of the sporophyte develop short branches, which accumulate starch and become tubers. If these are detached from the parent plant, they will develop into new sporophytes.

Equisetum arvense

DIVISION PTERIDOPHYTA

CLASS SPHENOPSIDA

ORDER EQUISETALES

GENUS *Equisetum*

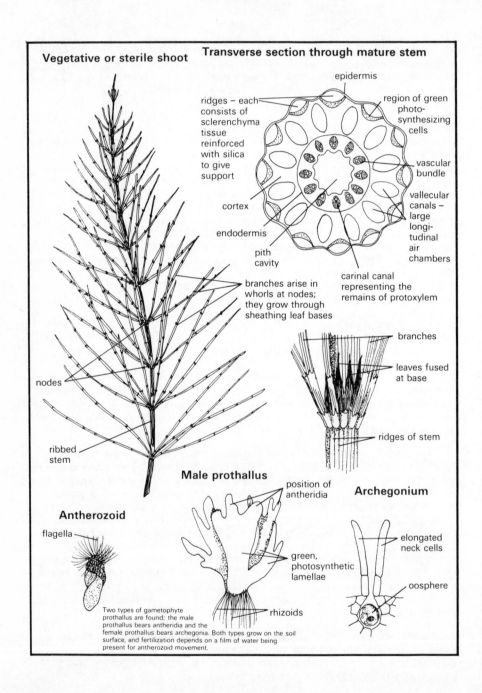

Vegetative or sterile shoot

Transverse section through mature stem

epidermis

ridges – each consists of sclerenchyma tissue reinforced with silica to give support

region of green photosynthesizing cells

vascular bundle

vallecular canals – large longitudinal air chambers

cortex

endodermis

pith cavity

carinal canal representing the remains of protoxylem

branches arise in whorls at nodes; they grow through sheathing leaf bases

branches

leaves fused at base

nodes

ridges of stem

ribbed stem

Male prothallus

position of antheridia

Archegonium

Antherozoid

flagella

green, photosynthetic lamellae

elongated neck cells

oosphere

rhizoids

Two types of gametophyte prothallus are found; the male prothallus bears antheridia and the female prothallus bears archegonia. Both types grow on the soil surface, and fertilization depends on a film of water being present for antherozoid movement.

42

Equisetum arvense

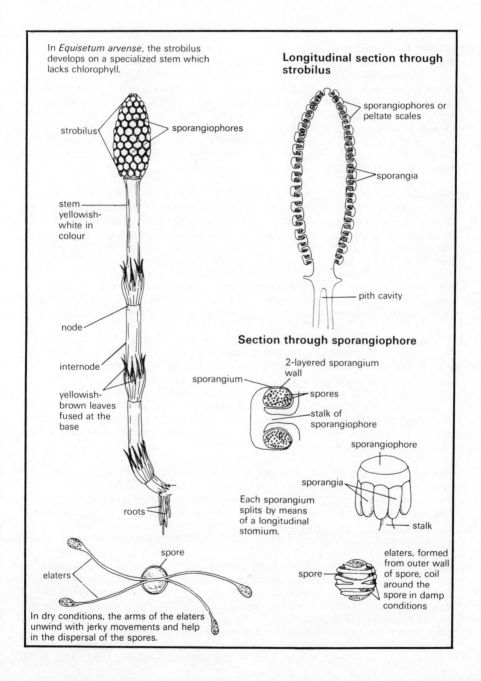

In *Equisetum arvense*, the strobilus develops on a specialized stem which lacks chlorophyll.

strobilus

sporangiophores

stem — yellowish-white in colour

node

internode

yellowish-brown leaves fused at the base

roots

elaters

spore

In dry conditions, the arms of the elaters unwind with jerky movements and help in the dispersal of the spores.

Longitudinal section through strobilus

sporangiophores or peltate scales

sporangia

pith cavity

Section through sporangiophore

sporangium

2-layered sporangium wall

spores

stalk of sporangiophore

sporangiophore

sporangia

Each sporangium splits by means of a longitudinal stomium.

stalk

spore

elaters, formed from outer wall of spore, coil around the spore in damp conditions

Class Pteropsida
Order Filicales

The members of this order are mostly terrestrial with some aquatic species. The Filicales have a world-wide distribution and, while most members are found in the humid warmth of tropical climates, all the common ferns of temperate regions are included in this order. A number of the tropical forms are epiphytic. A few genera, e.g. *Azolla* and *Salvinia*, are aquatic, growing either on the surface of the water or in the marshy edges of lakes and streams. The Filicales are of little economic importance, except as sources of medicines. An oil extracted from *Dryopteris* spp. has been used in the treatment of people infected with tapeworm. The extract paralyses the tapeworm and can cause some of it to be dislodged from the gut wall, but usually the scolex, or head region, remains firmly attached and continues to form more segments.

The sporophyte is usually macrophyllous. The gametophyte is small, green and photosynthetic. There is a wide range of form of the sporophyte generation in this order, but in the temperate genera the leaves are usually large, and develop at the apex of a creeping, underground rhizome which bears adventitious roots. The leaves, or fronds, in the majority of species are pinnate or deeply-divided. In many species, the fertile and sterile leaves are similar, e.g. *Dryopteris* spp., but in some the fertile leaves have a greatly reduced lamina, as in *Blechnum spicant*. Sporangia are borne on the fertile leaves, either on the undersurface in distinct groups called sori, or at the margins. The sorus is often covered by a protective structure, the indusium, which has a characteristic form in the different species.

The stem grows from an apical cell, and leaf primordia, which begin as slight bumps just behind the apex, arise in a definite sequence. Each leaf primordium soon develops its own apical cell. The mature vascular tissue, consisting of tracheids and sieve cells, shows some variation in its arrangement in the stele. The most common arrangement in this order is called a dictyostele and consists of an anastomosing system of vascular bundles with vascular strands going to the leaves. In addition to the vascular tissue, there is often sclerenchyma in rods or bands which contribute to the rigidity. In the leaves, there is differentiation of the lamina into palisade and mesophyll tissue.

The gametophyte may be filamentous, as in the filmy ferns, or it may be a flat, green, heart-shaped plate of cells, anchored to the soil by rhizoids arising from its undersurface, as in *Dryopteris* spp. The gametophytes, or prothalli, usually produce both male and female gametes in antheridia and archegonia respectively. The prothalli are short-lived, delicate structures, unable to withstand dry conditions, but totally independent of the sporophyte generation.

There is a heteromorphic alternation of generations in the life cycle. Sporangia develop from a single initial cell on the undersurface or margins of fertile leaves. The initial cell gives rise to a cluster of spore mother cells surrounded by a two-layered tapetum. The spore mother cells undergo meiosis and tetrads of haploid spores are formed. The spores are all the same. When mature, each sporangium has a stalk, a wall one cell thick and a well-defined region called the stomium. In the wall, in addition to the stomium, there is usually a band of cells with specially-thickened walls, the annulus. As the mature sporangium dries out, tensions are set up in the annulus, and the thin-walled cells of the stomium region eventually break. The upper part of the sporangium springs back, shooting out some of the spores. Later, the tension in the annulus is released and the upper part of the sporangium suddenly returns to its original position, jerking more spores out in the process. The spores germinate in moist conditions, growing into the gametophyte generation. Antheridia and archegonia develop on the undersurface of the same prothallus; the antheridia amongst the rhizoids and the archegonia in the region near the notch. Mature motile antherozoids are released from the antheridia and swim to the archegonia, where fertilization occurs. The neck of the archegonium secretes malic acid, to which the antherozoids are attracted. Antheridia mature and the antherozoids are released before the archegonia on the same prothallus are fully developed, so that it is unlikely that self-fertilization occurs very frequently. After fertilization of the oosphere, the diploid oospore develops into a sporophyte embryo, consisting of an absorptive foot, a stem apex, root and first leaf. The embryo is dependent on the prothallus in its initial stages of growth, but as the leaves develop, the prothallus tissue dies.

Hymenophyllum tunbridgense

DIVISION PTERIDOPHYTA

CLASS PTEROPSIDA

ORDER FILICALES

GENUS *Hymenophyllum*

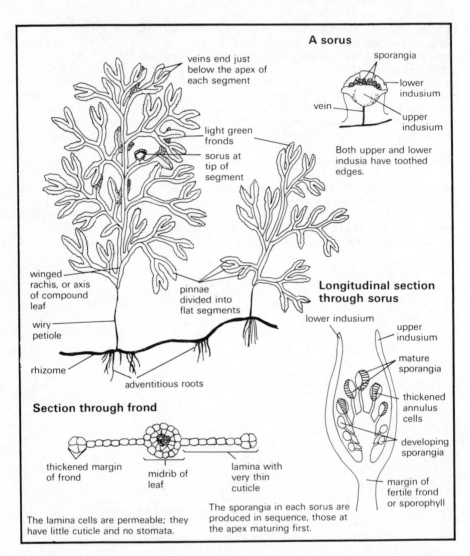

A sorus

veins end just below the apex of each segment

sporangia

lower indusium

vein

upper indusium

Both upper and lower indusia have toothed edges.

light green fronds

sorus at tip of segment

winged rachis, or axis of compound leaf

pinnae divided into flat segments

Longitudinal section through sorus

lower indusium

upper indusium

mature sporangia

thickened annulus cells

developing sporangia

margin of fertile frond or sporophyll

wiry petiole

rhizome

adventitious roots

Section through frond

thickened margin of frond

midrib of leaf

lamina with very thin cuticle

The lamina cells are permeable; they have little cuticle and no stomata.

The sporangia in each sorus are produced in sequence, those at the apex maturing first.

Hymenophyllum tunbridgense, Tunbridge filmy fern, is found growing in wet, shady habitats, such as the bases of tree trunks and on rocks by rivers or near water, in western parts of Britain. The spores ripen in June and July. The gametophyte thallus consists of an irregularly-branched green ribbon, bearing antheridia and archegonia on the margins of the ventral surface. The fronds of the sporophyte grow up to 10 cm long and they are light green in colour. The species is quite rare.

Dryopteris filix-mas

Dryopteris filix-mas, the male fern, grows in damp woods and hedgerows, in large clumps. The fronds are from 40 – 90 cm in length, and the spores ripen in July and August.

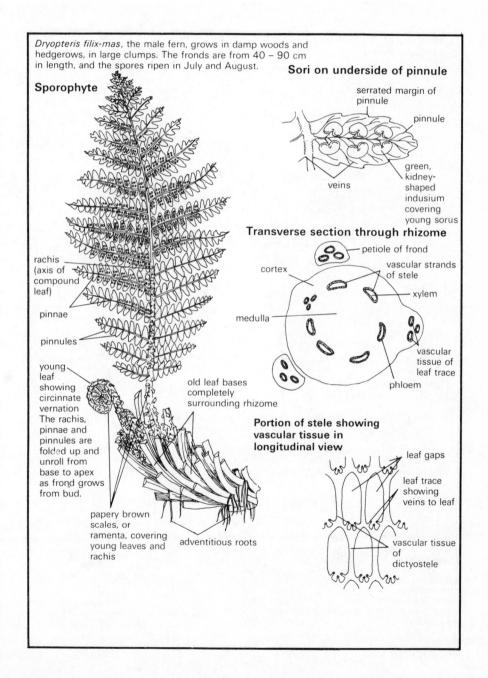

Sporophyte

Sori on underside of pinnule

serrated margin of pinnule

pinnule

green, kidney-shaped indusium covering young sorus

veins

rachis (axis of compound leaf)

pinnae

pinnules

young leaf showing circinnate vernation The rachis, pinnae and pinnules are folded up and unroll from base to apex as frond grows from bud.

old leaf bases completely surrounding rhizome

papery brown scales, or ramenta, covering young leaves and rachis

adventitious roots

Transverse section through rhizome

petiole of frond

cortex

vascular strands of stele

xylem

medulla

vascular tissue of leaf trace

phloem

Portion of stele showing vascular tissue in longitudinal view

leaf gaps

leaf trace showing veins to leaf

vascular tissue of dictyostele

Dryopteris filix-mas

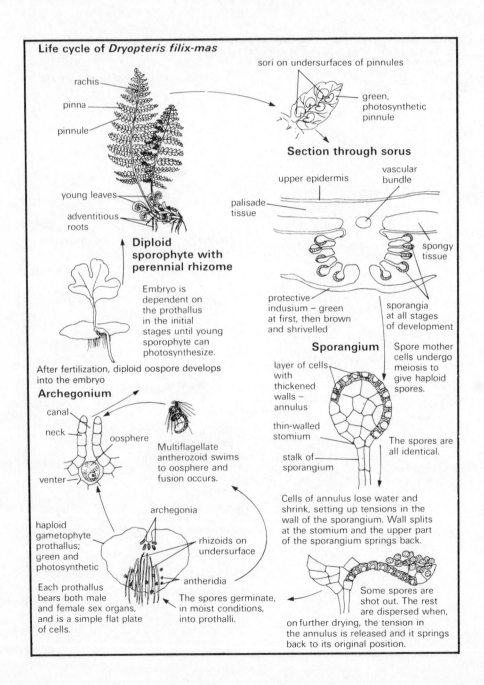

Life cycle of *Dryopteris filix-mas*

rachis

pinna

pinnule

sori on undersurfaces of pinnules

green, photosynthetic pinnule

young leaves

adventitious roots

Diploid sporophyte with perennial rhizome

Embryo is dependent on the prothallus in the initial stages until young sporophyte can photosynthesize.

After fertilization, diploid oospore develops into the embryo

Archegonium

canal

neck

oosphere

venter

Multiflagellate antherozoid swims to oosphere and fusion occurs.

archegonia

haploid gametophyte prothallus; green and photosynthetic

rhizoids on undersurface

antheridia

Each prothallus bears both male and female sex organs, and is a simple flat plate of cells.

The spores germinate, in moist conditions, into prothalli.

Section through sorus

upper epidermis

vascular bundle

palisade tissue

spongy tissue

protective indusium – green at first, then brown and shrivelled

sporangia at all stages of development

Sporangium

layer of cells with thickened walls – annulus

thin-walled stomium

stalk of sporangium

Spore mother cells undergo meiosis to give haploid spores.

The spores are all identical.

Cells of annulus lose water and shrink, setting up tensions in the wall of the sporangium. Wall splits at the stomium and the upper part of the sporangium springs back.

Some spores are shot out. The rest are dispersed when, on further drying, the tension in the annulus is released and it springs back to its original position.

Pteridium aquilinum

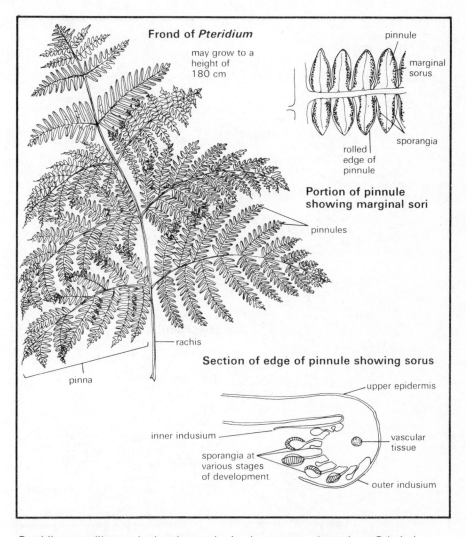

Frond of *Pteridium*

may grow to a height of 180 cm

pinnule

marginal sorus

sporangia

rolled edge of pinnule

Portion of pinnule showing marginal sori

pinnules

rachis

pinna

Section of edge of pinnule showing sorus

upper epidermis

inner indusium

vascular tissue

sporangia at various stages of development

outer indusium

Pteridium aquilinum, the bracken or brake, is common throughout Britain in woods, grassland and on heaths. It is a species which prefers acid soils, and is usually absent from chalk and limestone areas. It has very deep, widely-spreading rhizomes, and it appears to be able to compete with the flowering plants in its habitats, for growing space. *Pteridium* is poisonous to grazing animals if eaten in large quantities and, therefore, it is a nuisance to farmers. It is very difficult to eradicate because of its deep-rooted rhizomes, and tends to spread quite rapidly in pasture.

Some other Filicales

DIVISION PTERIDOPHYTA

CLASS PTEROPSIDA

ORDER FILICALES

GENUS *Phyllitis*
     Polypodium
     Asplenium
     Blechnum

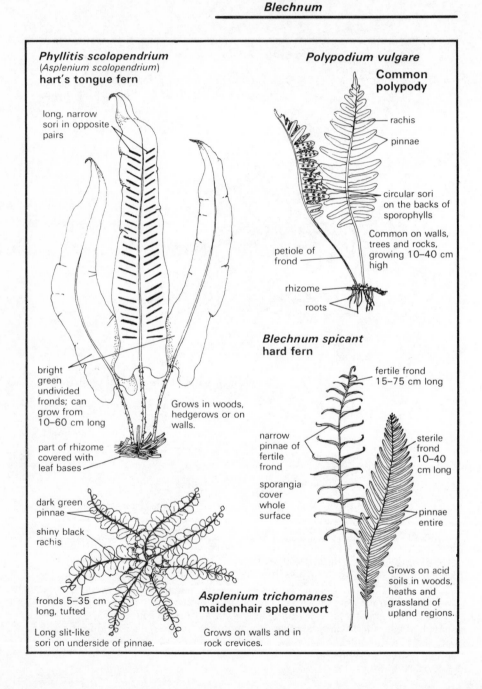

Phyllitis scolopendrium
(*Asplenium scolopendrium*)
hart's tongue fern

long, narrow sori in opposite pairs

bright green undivided fronds; can grow from 10–60 cm long

Grows in woods, hedgerows or on walls.

part of rhizome covered with leaf bases

dark green pinnae

shiny black rachis

fronds 5–35 cm long, tufted

Long slit-like sori on underside of pinnae.

Asplenium trichomanes
maidenhair spleenwort

Grows on walls and in rock crevices.

Polypodium vulgare
Common polypody

rachis

pinnae

circular sori on the backs of sporophylls

Common on walls, trees and rocks, growing 10–40 cm high

petiole of frond

rhizome

roots

Blechnum spicant
hard fern

fertile frond 15–75 cm long

narrow pinnae of fertile frond

sporangia cover whole surface

sterile frond 10–40 cm long

pinnae entire

Grows on acid soils in woods, heaths and grassland of upland regions.

Division *Spermatophyta*

Mostly terrestrial; some aquatic species.
> This is the most widespread and successful division of plants. The Spermatophyta have colonized nearly all the habitats, and now form the dominant vegetation of the world. Within the division there is a wide range of form; the more advanced groups contain herbaceous plants as well as trees and shrubs. Many of the plants are of economic importance, providing food, building materials and drugs.

The sporophyte is the persisting plant body, always showing differentiation into stem, root and leaf systems; gametophytes are very reduced and never independent.
> The Spermatophyta do not often show dichotomous branching, but there is always a high degree of cell specialization and tissue differentiation. The xylem is composed of tracheids, and vessels occur in the more advanced types. Growth of the roots and the stems occurs from an apical meristem and lateral meristems, whose activities result in secondary thickening, are common. The male and female gametophytes are reduced structures; the male gametophyte is contained within the pollen grain and the female gametophyte remains within the megaspore, which is never shed from the sporophyte.

Autotrophic nutrition; photosynthetic pigments are present.
> Chlorophylls a and b, together with beta-carotene and xanthophylls are present in chloroplasts enabling photosynthesis to occur. One or two parasitic species lack chlorophyll and derive their nutrition from their host, as in *Cuscuta* sp. (dodder) and *Orobanche* sp. (broomrape). Some angiosperms have become parasitic, but have not lost their chlorophyll. Such is the case with *Viscum* sp. (mistletoe) and *Euphrasia* sp. (eyebright).

Heteromorphic alternation of generations in the life cycle.
> The sporophyte bears specialized reproductive structures, which are cones or strobili in the lower groups and flowers in the more advanced groups. Two kinds of spores are produced; microspores, which become pollen grains, and megaspores, which develop in the ovules and give rise to the female gametophytes. In this division, the megasporangium wall has given rise to protective coverings, or integuments, and a nutritive tissue called the nucellus has evolved. The megaspores are embedded in the nucellus and the female gametophytes which develop are never exposed. This unit, which consists of integuments, nucellus and female gametophyte, is known as an ovule. Antheridia and archegonia are produced in the primitive types. The pollen grains are shed from the sporophyte and transferred to the female structures by air currents or insects (pollination). Prior to fertilization, the pollen grain germinates on the female structure, producing a pollen tube, which penetrates the female tissues. The male nuclei, or gametes, pass down the pollen tube to the female nucleus, thus eliminating the need for water to be present. After fertilization, seeds are formed. Each seed contains an embryo sporophyte, together with a food store and surrounded by a seed coat, or testa. After being shed from the parent sporophyte, the seed can remain dormant,

surviving unfavourable conditions, and eventually germinates to give rise to a new sporophyte. The energy and food materials needed for successful germination come from the food reserves of the seed, which were synthesized by the parent sporophyte.

The evolution of the Spermatophyta probably took place in the Devonian period of the Palaeozoic era, as fossil forms have been found from the rock strata dating from that time. It is very likely that the seed-bearing plants evolved from the early fern-like plants, but despite fossil Spermatophyta having been found alongside fossils of the ferns and their allies, there is not yet any direct evidence for this theory.

There are three classes in this division:
    Class Pteridospermae the seed-bearing ferns.
                            These plants are all extinct and known only as fossils.
                            They had fern-like leaves and produced seeds.
    Class Gymnospermae the naked-seeded plants.
    Class Angiospermae the flowering plants with enclosed seeds.

Class Gymnospermae
Naked-seeded plants

The members of this class are all terrestrial and range in size from small shrubs, such as juniper, to very large trees, such as the giant redwoods of California. The gymnosperms have a worldwide distribution with the greatest number of species occurring in the northern temperate regions. Many of these northern species can tolerate the coldest conditions and highest altitudes at which trees grow. Members of this class are of great economic importance in supplying valuable timber for building and furniture, wood pulp for the paper industry, resins and turpentine.

The sporophyte plant is always a woody perennial shrub or tree showing differentiation into stem, leaf and root systems. There is secondary vascular tissue present; the xylem typically consists of tracheids and the phloem contains sieve cells.

The reproductive structures borne on the sporophyte are arranged in cones or strobili, which are unisexual. The male cones consist of microsporophylls bearing microsporangia, in which microspores are formed. The male gametophyte is much reduced; it consists of one or two prothallial cells and an antheridial cell. The female cones consist of megasporophylls bearing megasporangia or ovules. Each ovule is made up of a megaspore enclosed in the nucellus of the sporophyte, which is surrounded by an integument. The female gametophyte develops in the megasporangium and is a fairly large multicellular structure which bears archegonia. It is always retained on the sporophyte plant within the megasporangium. Microspores from the male cones are carried by wind and air currents to the female cones and pollination is effected, after which the microspores germinate. A pollen tube is produced which grows towards the female archegonia, and fertilization occurs when one male nucleus fuses with the female nucleus of an oosphere. Seed development follows fertilization. Each seed consists of an embryo, which develops from the diploid zygote, a food store derived from the female prothallus, surrounded by a testa formed from the integument of the ovule. The seeds are borne on the surface of the megasporophylls, unlike the Angiospermae where the seeds are enclosed in special structures called carpels.

There are a number of orders recognized in this class, some of which are only represented by fossil types, and others which contain both living and fossil representatives.
Examples from two of these orders are considered in further detail:
    Order Cycadales   *Cycas revoluta*
    Order Coniferales   *Pinus sylvestris*

Order Cycadales

This is a small group of terrestrial plants, found in tropical climates in Africa, eastern Asia and Australia. The group came into prominence in the Mesozoic era and the representatives which survive today display many of the more primitive characteristics of the seed-bearing plants. The members of this order are not as hardy as the Coniferales and cannot compete successfully with the flowering plants in temperate climates. The pith of the sago palm (*Cycas revoluta*) contains large amounts of starch, and is important commercially as one of the sources of sago.

The sporophytes are upright; with thick, unbranched stems, large pinnate leaves and large tap roots. In this order, the stems are either short and stocky with a large portion below the ground or much taller, growing to a height of about 15 metres. Growth of the stem is meristematic and very slow and secondary thickening occurs. Two types of leaves are produced; foliage leaves which are large and compound pinnate, and scale leaves which cover the apex and the upper part of the stem. The foliage leaves form a crown at the apex of the stem. The massive tap roots bear some normal lateral roots together with other roots occurring at the soil surface and containing fungi and blue-green algae in their tissues.

There is a great deal of parenchyma tissue in the stele of the stem and the main mechanical support for the stem is given by the leaf bases, which contain much sclerenchyma tissue. The old leaf bases persist and surround the lower parts of the stem, making it fairly thick. The foliage leaves have a very definite cuticle and clearly differentiated palisade and mesophyll tissue.

In this order, male and female strobili are borne on separate plants (dioecious), and they occur either at the apex of the sporophyte or develop laterally.

The male cones consist of a large number of microsporophylls arranged spirally. The microsporophylls are woody, wedge-shaped structures bearing large numbers of microsporangia on the lower surface. In some species; the microsporangia appear to be grouped together in sori. Inside the microsporangia, microspore mother cells undergo meiosis to give rise to tetrads of haploid microspores, which are released on dehiscence of the microsporangium. Female cones vary more than the male cones. In some species, the megasporophylls are loosely grouped together and bear several pairs of ovules, while in other species the megasporophylls are tightly compacted together, each megasporophyll bearing only two ovules. In all species, however, the female structures are very large, and in *Macrozamia* the ovules may reach a length of 6 cm. Wind pollination occurs and the pollen grains (microspores) are trapped in a drop of mucilage, which is secreted at the micropylar end of the ovule. The microspores are drawn in to the ovule as the mucilage dries. They begin to germinate, the pollen tubes growing into part of the ovule tissue, the nucellus, which is broken down and digested by enzymes. The nucellus tissue is disorganized and a clear passage forms above the archegonia which have developed on the female prothallus. In each male gametophyte, two coiled, multiflagellate antherozoids develop and are released into the space above the archegonia when the pollen tubes burst. The antherozoids pass into the archegonia and down the necks. One or more of the oospheres may be fertilized. Only one viable embryo will eventually develop. There may be an interval of some months between pollination and fertilization, but once fertilization has occurred, the embryo begins to develop straight away, deriving its nourishment from the female gametophyte tissue. The embryo, consisting of two or more cotyledons, a stem apex (plumule) and a hypocotyl, is surrounded by a hard, stony integument, which also encloses the remains of the female prothallus. The seeds germinate straight away when they have been shed, provided that conditions are suitable. The seeds do not undergo a period of dormancy and do not remain viable for more than a few months. When germination does occur, the hypocotyl develops a tap root, the plumule emerges and the cotyledons remain partly enclosed by the seed coat.

Cycas revoluta

DIVISION SPERMATOPHYTA

CLASS GYMNOSPERMAE

ORDER CYCADALES

GENUS *Cycas*

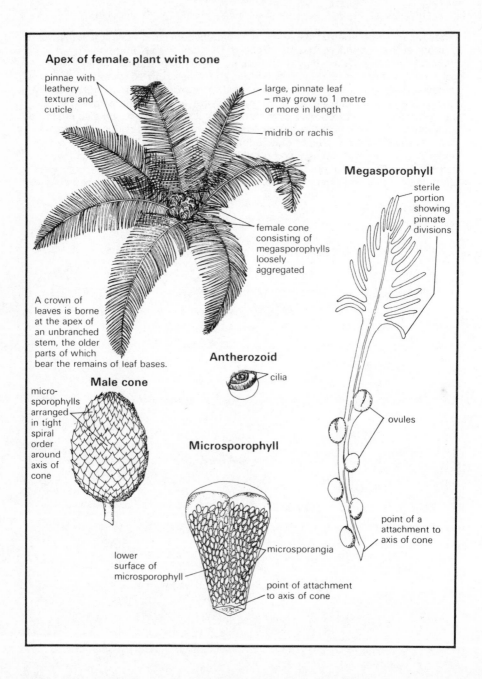

Apex of female plant with cone

pinnae with leathery texture and cuticle

large, pinnate leaf – may grow to 1 metre or more in length

midrib or rachis

Megasporophyll

sterile portion showing pinnate divisions

female cone consisting of megasporophylls loosely aggregated

A crown of leaves is borne at the apex of an unbranched stem, the older parts of which bear the remains of leaf bases.

Antherozoid

cilia

Male cone

micro-sporophylls arranged in tight spiral order around axis of cone

ovules

Microsporophyll

microsporangia

lower surface of microsporophyll

point of attachment to axis of cone

point of attachment to axis of cone

Cycas revoluta

Life cycle of *Cycas revoluta*

Diploid sporophyte

Dioecious; male and female reproductive structures borne on separate plants.

↓

Male cones made up of microsporophylls bearing large numbers of microsporangia on the lower surface.

↓

Microspore mother cells undergo meiosis to give haploid microspores, or pollen grains.

prothallial cell

antheridial cell — spore wall

tube nucleus

Microspores are shed from the male cone.

↓

Pollination occurs. The microspores are tiny and easily borne in air currents. The ovule exudes a mucilaginous 'pollination drop' in which the pollen grains are caught and drawn into the pollen chamber.

↓

Pollen germinates; the pollen tube grows into the nucellus tissue at the base of the pollen chamber. The nucellus tissue is digested and used for the nutrition of the pollen tube.

prothallial cell

developing antherozoids

pollen tube — tube nucleus

Female cones made up of megasporophylls bearing ovules. Each ovule consists of a nucleus almost completely surrounded by integument except for a small gap, the micropyle.

↓

Archesporial cell, or megaspore mother cell, develops in the nucellus and undergoes meiosis to give 4 haploid potential megaspores; 3 abort.

↓

Remaining megaspore increases in size, and develops into the female prothallus. Some of the nucellus tissue at the micropylar end breaks down to form the pollen chamber.

↓

micropyle

fleshy layers

pollen chamber

stony layer

nucellus

female prothallus containing reserves of starch

archegonial chamber

archegonium

↓

The nucellus tissue below the pollen chamber breaks down, providing a passage for the antherozoids to reach the archegonia.

The pollen tubes grow larger and hang down into the archegonial chamber.

The pollen tube bursts, and the antherozoids are released. They swim to the archegonia.

antherozoid

neck cells

oosphere

archegonium

At fertilization, an antherozoid, which comes into contact with the neck of an archegonium, gets sucked down to the oosphere. The cilia are shed and the two nuclei fuse.

A diploid zygote is formed and begins to divide to give rise to the embryo, using food from the female prothallus. The mature seed contains 1 or 2 cotyledons enclosing a plumule, and a hypocotyl. The seeds are shed and germination occurs straight away if conditions are suitable.

Order Coniferales

The conifers are the most widely-distributed of the gymnosperms, and play a dominant role in the vegetation of the colder parts of the temperate zones and at high altitudes. Conifers are not as common in the tropical regions. They are economically important in providing timber for building and papermaking, and some species produce considerable quantities of resin, which is a valuable source of turpentine.

The sporophytes are mostly trees or shrubs; with stems showing monopodial growth, shallow roots and leaves showing a variety of form, but are mostly needle-like. Because of the monopodial growth, where the apical bud continues to grow each year, the shape of the trees often resembles a cone or pyramid. Many species are long-lived and can attain heights of more than 100 metres. The roots of the conifers do not penetrate very deeply into the soil, and many species have mycorrhizal associations with fungi. The leaves may be small and needle-like, as in *Pinus* spp., or small scales as in *Cupressus* spp. In *Araucaria* spp., the leaves are broad, reaching 5 cm or more in length. In some species, the leaves are borne on short, or dwarf, shoots and in nearly all species, the leaves persist for several seasons.

The stems grow from groups of meristematic cells, and, when mature, consist mainly of secondary wood in which it is easy to see annual rings. These rings are caused by a difference in size of the tracheids formed during the course of each growing season. The tracheids formed at the end of the season are narrower and thicker-walled than those formed at the beginning. Resin canals occur and are found running longitudinally in the xylem. The resin is made by the epithelial cells of the canals; it is a complex substance containing phenols and terpenes.

Most leaves have a thick cuticle and a hypodermis, which may be lignified, just below the epidermis. Resin canals may also be present in the leaves.

In most conifers, male and female cones are produced in different regions on the same tree. In *Pinus* spp., female cones develop near the apex of the tree and male cones are produced on the lower branches. A few species are dioecious, e.g. *Juniperus* sp. (juniper). The male cones are uniform in structure, consisting of microsporophylls, each bearing two microsporangia on the undersurface, borne spirally on a central axis. Inside each microsporangium, microspore mother cells undergo meiosis to give rise to haploid microspores, or pollen grains. Each pollen grain eventually has a three-layered wall and, in many species, wings or air bladders to help in the dispersal by air currents. The female cones vary in structure according to the species, but generally consist of a central axis bearing spirally arranged bract scales. In the axils of these bract scales, ovuliferous scales develop and become much larger and woody, so that in many species it is not possible to see the bract scales in the mature cone. Usually two ovules develop on the surface of each ovuliferous scale.

An ovule consists of an inner mass of cells called the nucellus and an outer layer of cells, the integument. The integument does not entirely enclose the nucellus; there is a small pore, or micropyle, at one end. Four potential haploid megaspores are formed, of which only one undergoes any further development. At pollination, the ovule contains one haploid megaspore. Wind pollination occurs and is similar to that already described for the cycads. In the conifers, the female cone elongates, forcing the scales apart, thus allowing the pollen grains to enter. Very shortly after pollination, the female cone grows again and the scales close up tightly once more.

Germination and growth of the pollen grain are slow. During the first season, growth of the pollen tube begins, but ceases during the winter. The following spring, the female gametophyte develops from the megaspore and becomes cellular, with 1 to 6 archegonia arising at the micropylar end. The pollen tube continues to grow towards the archegonia, and fertilization occurs when the pollen tube penetrates the archegonium and the two male nuclei are released. Only one nucleus fuses with the oosphere nucleus, the other male nucleus disintegrating. Fertilization occurs during the second season, during which time the female cone gets bigger, but remains tightly closed. Immediately after fertilization, the diploid zygote begins to grow into an embryo. Despite the fact that several oospheres may have been fertilized, only one embryo will mature; the rest abort. Each embryo consists of a plumule, a radicle and from 5 to 10 cotyledons. The energy and food for this development comes from the female prothallus. The integument of the ovule becomes hard and forms the seed coat, enclosing the embryo together with what is left of the nutritive female prothallus. The female cone becomes harder and woody during the third season, and the scales open up, exposing the ripe seeds on the surface of each ovuliferous scale. Each seed has a thin papery wing attached to it. It soon becomes free of the scale, falls out of the cone and is dispersed by air currents.

Germination occurs in favourable conditions. The radicle emerges, breaking through the seed coat and anchoring the young plant in the soil. The hypocotyl elongates, carrying the cotyledons and stem apex above the soil. The cotyledons soon become green and begin to photosynthesize.

Pinus sylvestris

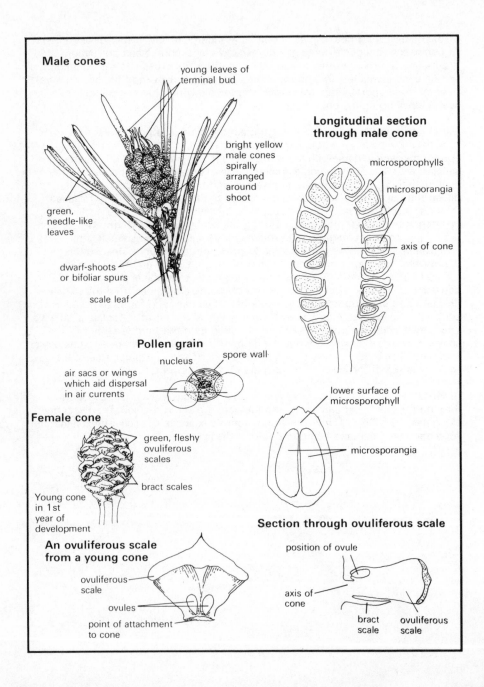

Male cones

young leaves of terminal bud

bright yellow male cones spirally arranged around shoot

green, needle-like leaves

dwarf-shoots or bifoliar spurs

scale leaf

Longitudinal section through male cone

microsporophylls

microsporangia

axis of cone

Pollen grain

nucleus

spore wall

air sacs or wings which aid dispersal in air currents

lower surface of microsporophyll

microsporangia

Female cone

green, fleshy ovuliferous scales

bract scales

Young cone in 1st year of development

Section through ovuliferous scale

position of ovule

An ovuliferous scale from a young cone

ovuliferous scale

ovules

point of attachment to cone

axis of cone

bract scale

ovuliferous scale

Pinus sylvestris

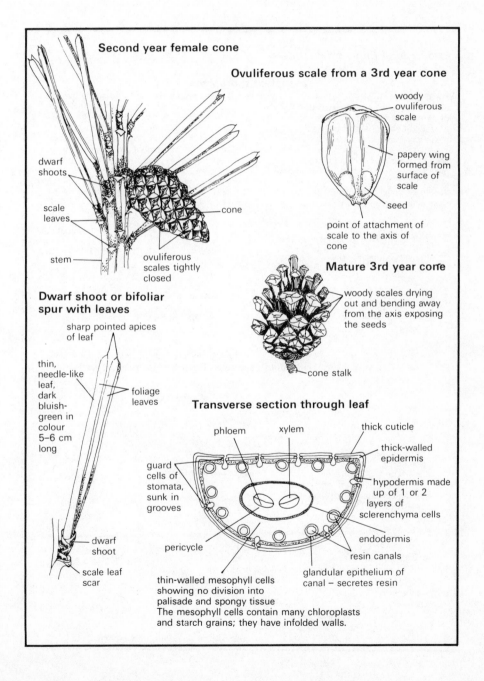

Second year female cone

dwarf shoots

scale leaves

stem

ovuliferous scales tightly closed

cone

Ovuliferous scale from a 3rd year cone

woody ovuliferous scale

papery wing formed from surface of scale

seed

point of attachment of scale to the axis of cone

Mature 3rd year cone

woody scales drying out and bending away from the axis exposing the seeds

cone stalk

Dwarf shoot or bifoliar spur with leaves

sharp pointed apices of leaf

thin, needle-like leaf, dark bluish-green in colour 5–6 cm long

foliage leaves

dwarf shoot

scale leaf scar

Transverse section through leaf

phloem

xylem

thick cuticle

thick-walled epidermis

guard cells of stomata, sunk in grooves

hypodermis made up of 1 or 2 layers of sclerenchyma cells

endodermis

resin canals

pericycle

glandular epithelium of canal – secretes resin

thin-walled mesophyll cells showing no division into palisade and spongy tissue
The mesophyll cells contain many chloroplasts and starch grains; they have infolded walls.

Pinus sylvestris

Life cycle of *Pinus sylvestris*

Diploid sporophyte

First year:

Male cones arise in clusters round the bases of terminal buds. Each cone is made up of microsporophylls, each bearing 2 microsporangia on the lower surface

↓

Microspore mother cells undergo meiosis to give haploid microspores, or pollen grains.

When the pollen grains are shed from the male cone, each one contains

↓

2 small prothallial cells, an anthiderial cell and a tube nucleus.

↓

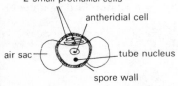

2 small prothallial cells

antheridial cell

air sac — — tube nucleus

spore wall

Female cones are produced laterally in the axils of scale leaves on young branches; usually only 1 or 2 per branch. The cones are made up of megasporophylls, which consist of an ovuliferous scale and a bract scale, with two ovules per scale on the upper surface.

Each ovule consists of the nucellus, in which the archesporial cell, or megaspore mother cell undergoes meiosis.

↓

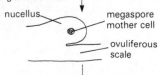

nucellus — — megaspore mother cell

— ovuliferous scale

↓

4 haploid megaspores are produced; 3 abort. The surviving megaspore develops into the female prothallus.

At pollination, the edges of the scales of the female cone roll inwards and the scales are forced apart, allowing pollen grains to enter. The pollen grains are trapped in the mucilage at the micropyle.

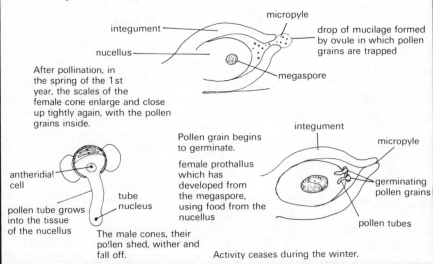

micropyle

integument —

drop of mucilage formed by ovule in which pollen grains are trapped

nucellus —

megaspore

After pollination, in the spring of the 1st year, the scales of the female cone enlarge and close up tightly again, with the pollen grains inside.

Pollen grain begins to germinate.

integument

micropyle

female prothallus which has developed from the megaspore, using food from the nucellus

antheridial cell

tube nucleus

pollen tube grows into the tissue of the nucellus

germinating pollen grains

pollen tubes

The male cones, their pollen shed, wither and fall off.

Activity ceases during the winter.

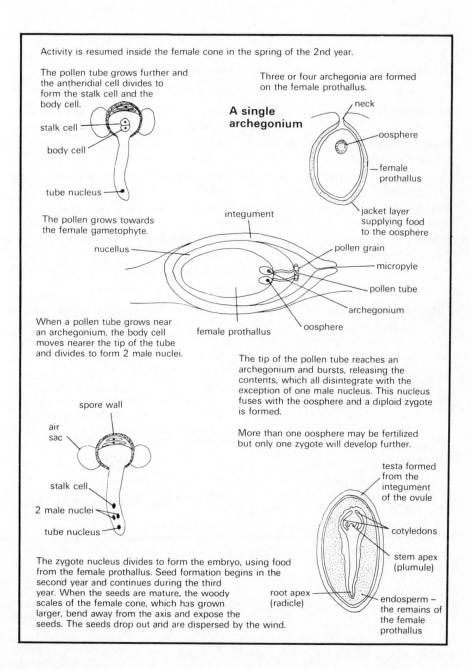

Activity is resumed inside the female cone in the spring of the 2nd year.

The pollen tube grows further and the antheridial cell divides to form the stalk cell and the body cell.

stalk cell

body cell

tube nucleus

Three or four archegonia are formed on the female prothallus.

A single archegonium

neck

oosphere

female prothallus

jacket layer supplying food to the oosphere

The pollen grows towards the female gametophyte.

integument

nucellus

pollen grain

micropyle

pollen tube

archegonium

oosphere

female prothallus

When a pollen tube grows near an archegonium, the body cell moves nearer the tip of the tube and divides to form 2 male nuclei.

The tip of the pollen tube reaches an archegonium and bursts, releasing the contents, which all disintegrate with the exception of one male nucleus. This nucleus fuses with the oosphere and a diploid zygote is formed.

More than one oosphere may be fertilized but only one zygote will develop further.

spore wall

air sac

stalk cell

2 male nuclei

tube nucleus

testa formed from the integument of the ovule

cotyledons

stem apex (plumule)

The zygote nucleus divides to form the embryo, using food from the female prothallus. Seed formation begins in the second year and continues during the third year. When the seeds are mature, the woody scales of the female cone, which has grown larger, bend away from the axis and expose the seeds. The seeds drop out and are dispersed by the wind.

root apex (radicle)

endosperm – the remains of the female prothallus

Summary of differences between gymnosperms and angiosperms

Gymnosperms	Angiosperms
Mostly trees and shrubs.	Herbaceous plants as well as trees and shrubs.
Worldwide distribution, but not many different forms.	Worldwide distribution, with a great variety of different forms.
Xylem composed of tracheids only.	Xylem composed of vessels and tracheids.
Reproductive parts borne on sporophylls aggregated into cones or strobili.	Reproductive parts in specialized structures called flowers, consisting of fertile and sterile whorls.
Cones or strobili unisexual.	Flowers usually hermaphrodite, containing male and female parts.
Wind pollination only.	Wind or insect pollination.
Female gametophyte is multicellular.	Female gametophyte represented by the embryo sac, a vacuolated cell containing 8 nuclei.
Archegonia formed in ovule.	No archegonia are formed.
At fertilization, one male nucleus fuses with the female oosphere nucleus: the second male nucleus breaks down.	At fertilization, one male nucleus fuses with the female nucleus; the second male nucleus fuses with 2 other nuclei (polar nuclei) in the embryo sac to give a 3n endosperm nucleus. This is known as double fertilization.
Seeds have one integument.	Seeds have two integuments.
Ovules develop on the surface of ovuliferous scales; not enclosed.	Ovules develop within a closed structure, the carpel.

Class Angiospermae
Plants with enclosed seeds

The members of this class form the dominant vegetation in the world today and are mainly terrestrial with some aquatic representatives. They are a widespread group showing great variety of growth form in the sporophyte. The sporophytes may be herbaceous, with aerial parts dying down during unfavourable periods, or woody, with persistent aerial stems protected by an outer covering of bark. Amongst the herbaceous types are the annuals that survive unfavourable periods as embryos enclosed in seeds, and the perennials producing flowers and seeds annually and resting buds associated with a perennating organ containing a food store. The angiosperms have evolved the most adaptations to successful life on land, and are represented in practically every habitat where life is possible. The aquatic representatives are mostly fresh-water species; those capable of tolerating marine conditions being much rarer. As might be expected, this group is of considerable economic importance to humans, providing food crops, pasture for domestic animals, fibres for clothing and textiles, materials for building, and many drugs and medicines.

The sporophyte plant is usually differentiated into stem, leaf and root systems, which show an almost unlimited variety of different forms, all based on a common plan. The main stem is usually upright, branching occurs in the axils of leaves and may develop in two ways. If the apical bud continues to grow indefinitely and remain permanently active, branching is monopodial; the main axis assumes the dominant position and the development of axillary buds is suppressed. This type of branching is common in the gymnosperms and, amongst the angiosperms, is typical of *Fraxinus* sp. (ash) and *Acer* sp. (sycamore). If, however, the apical bud gives rise to an inflorescence, any new growth in length of the stem must come from axillary buds, giving rise to sympodial branching, as seen in *Tilia* sp. (lime) and *Syringa* sp. (lilac). In some angiosperms, the stem grows horizontally forming a rhizome, as in *Iris* spp., or is very condensed as in the bulbs and corms of *Narcissus* sp. and *Crocus* sp. respectively, or may be especially adapted for the storage of food as in the tubers of *Solanum tuberosum* (potato).

Angiosperm leaves show an enormous range of size and shape. They usually consist of a flat blade, the lamina, and are attached to the stem, at a node, by a leaf stalk, the petiole. That portion of the stem between two leaf insertions is known as the internode. Often there are lateral, leaf-like outgrowths at the base of the petiole known as stipules, e.g. *Rosa* spp. Leaves are either simple, with all parts of the lamina continuous, or compound, made up of several leaflets joined to a common petiole. Simple leaves may be divided into lobes and notched. Compound leaves may have their leaflets arranged pinnately, in two rows either side of the petiole, e.g. *Pisum* spp., or palmately, where the leaflets radiate from a central point, e.g. *Lupinus* sp. (lupin) and *Aesculus* sp. (horse chestnut).

The roots of angiosperms show similar internal organization to those of the gymnosperms. Two types of root systems are found, the fibrous root system and the tap root system. A fibrous root system develops either from the profuse branching of the primary root, as in *Senecio* sp. (groundsel), or if the primary root dies and other roots develop adventitiously from the base of the stem, as is the case in members of the Graminae. If the primary root becomes established and develops lateral roots, a tap root system is formed. Some angiosperms produce tap roots which accumulate food stores and become swollen. These roots overwinter with dormant buds, and produce flowers and seeds the following season. These tap roots are useful food plants for humans and domestic animals, familiar examples being parsnips, carrots and beets.

The roots and stems of many angiosperms undergo secondary thickening and the vascular tissue is distinguished from that of the gymnosperms by the presence of vessels, as well as tracheids, in the xylem, and sieve tubes and companion cells in the phloem. Vessels are long tubes, which are made up of vessel segments, each derived from a single cell. During the formation of the vessel, the end walls of the segments break down leaving perforation plates. Sieve tubes are derived from rows of sieve cells arranged end-to-end, the end walls developing sieve plates which enable the passage of materials from cell to cell. Companion cells arise from the same mother cell as the sieve cells, but are smaller, have dense cytoplasm and a prominent nucleus. Growth in stems and roots is from groups of meristematic cells at the apices, but growth of most leaves is from initial cells at the margins followed by cell expansion. However, in the grasses, there is a meristem at the base of the lamina which continues to produce new cells until the leaf reaches its mature length.

The reproductive structures characteristic of the angiosperms are flowers, either borne singly or in clusters. The arrangement of the flowers is termed the inflorescence of the plant, and several types are distinguished, often characteristic of the families in which they occur. Most flowers are hermaphrodite, the stamens forming the androecium in which the male gametes develop, and the carpels forming the gynaecium which produce the female gametes. In addition to these essential parts of the flower, there are often associated structures, such as sepals and petals protecting the developing gametes or serving to attract the pollinating insects to the plants. Each stamen consists of a stalk or filament which bears a lobed anther, containing 2 or 4 pollen sacs at its apex. The stamen can be compared with a microsporophyll bearing microsporangia (the pollen sacs) inside which the microspores (pollen grains) are formed. Development of the microspores is very similar to that found in the gymnosperms. Each carpel is a closed structure, inside which one or more ovules develop. Each ovule develops from the nucellus and becomes surrounded by two integuments. The lower part of the carpel, which surrounds the ovules, is called the ovary and bears at its apex a sterile portion, the style, which terminates in a receptive surface, the stigma. It is convenient to think of a carpel as having evolved from a flat megasporophyll, which has become folded and joined along its edges to form a hollow container. Megaspores which develop in the ovules are thus completely enclosed, distinguishing the angiosperms from the gymnosperms.

When microspores, or pollen grains, are shed from the pollen sacs, they possess a generative nucleus, representing the male gametophyte, which later divides to produce two male gametes, and a pollen tube nucleus. The male gametophyte is thus even more reduced than in the gymnosperms, there being no vegetative cells present at all. There is similar reduction in the female gametophyte which develops within the ovule. A meiotic division of the megaspore mother cell results in the fomation of four potential haploid megaspores. Three of the four abort and only the inner one develops further. It enlarges and undergoes three successive mitotic divisions resulting in the formation of a large, vacuolated embryo sac containing eight nuclei characteristically arranged. Three of the nuclei are grouped at the micropyle end of the ovule, the central one being the female nucleus. At the opposite end is a group of three more nuclei, and two nuclei, the polar nuclei, are situated in the centre of the embryo sac. No archegonia are formed. After successful pollination, by wind or insects depending on the type of flower, the microspore germinates on the stigma, producing a pollen tube which grows through the sterile tissue of the carpel until it reaches the ovule. Fertilization occurs when the pollen tube reaches the micropyle and embryo sac, bursts, releases its two male gametes, one of which fuses with the female nucleus to form the diploid zygote, and the other fuses with the two polar nuclei to give rise to a triploid (3N) nucleus from which the endosperm is derived. This constitutes the double fertilization characteristic of the angiosperms. The diploid zygote develops into an embryo, consisting of a plumule, a radicle and one or two cotyledons. It is surrounded by the nutritive endosperm tissue and protected by the integuments which become the seed coat, or testa. The ovary undergoes changes and matures into a true fruit, which is often involved in the dispersal of the seeds. The ripe seeds usually germinate into new plants following a period of dormancy.

Asexual reproduction is shown in this group and either occurs by a process known as apomixis, or by purely vegetative means. Parthenogenesis is one form of apomixis, in which the egg develops without fertilization having taken place. The genetics of such a form of reproduction is complicated and it is interesting to note that the stimulus of pollination is often needed before parthenogenesis will occur. This type of reproduction can occur in *Taraxacum* sp. (dandelion). In *Festuca vivipara*, the formation of dwarf shoots, or bulbils, occurs on the inflorescence, instead of flowers. Amongst the purely vegetative forms of asexual reproduction, the rooting of stems, such as *Rubus* sp. (blackberry), the formation of plantlets on the leaf margins of *Kalanchoe* sp., the development of daughter bulbs and corms from the axillary buds of *Narcissus* sp. and *Crocus* sp., and the formation of buds on fleshy roots are common examples. In addition, gardeners and horticulturists propogate plants by means of layering, grafting and budding, thus ensuring the survival of useful and productive varieties.

This class is sub-divided into
  Sub-class Monocotyledones   mostly herbaceous plants
  Sub-class Dicotyledones   herbs, trees and shrubs.

A comparison of the Sub-classes Monocotyledones and Dicotyledones

Monocotyledones	Dicotyledones
Mostly herbaceous; only a few species showing the palm habit, but no true trees.	Herbs, shrubs and trees.
Adventitious roots present giving rise to fibrous root systems.	Tap root systems as well as fibrous root systems.
Stems have random distribution of vascular bundles.	Vascular bundles in the stem arranged in one or more rings.
Secondary growth rare.	Secondary growth common.
Leaves usually show parallel venation and are often isobilateral.	Leaves frequently show reticulate venation and are more usually dorsiventral.
Flower parts in 3s or multiples of 3.	Flower parts in 4s or 5s, or multiples of these numbers.
Sterile whorls of perianth segments in the flower; no distinction between petals and sepals.	Sterile whorls in the flower distinguished into petals and sepals.

One cotyledon in the seed.
  Examples:
Order Liliflorae
  Family Liliaceae
  Family Iridaceae
  Family Amaryllidaceae
Order Glumiflorae
  Family Graminae
	Two cotyledons in the seed.
	  Examples:
	Archichlamydeae petals free or absent
	Order Ranales
	  Family Ranunculaceae
	Order Rhoeadales
	  Family Cruciferae
	Order Rosales
	  Family Leguminosae
	      (Papilionaceae)
	Family Rosaceae
	Metachlyamydeae petals united
	Order Tubiflorae
	  Family Scrophulariaceae
	  Family Labiatae
	Order Asterales
	  Family Compositae

General structure of the Angiospermae

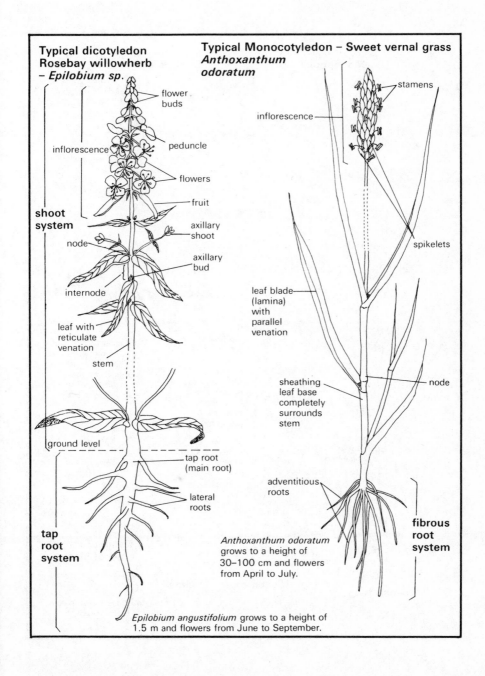

Typical dicotyledon
Rosebay willowherb
– *Epilobium sp.*

- flower buds
- inflorescence
- peduncle
- flowers
- fruit
- **shoot system**
- axillary shoot
- node
- axillary bud
- internode
- leaf with reticulate venation
- stem
- ground level
- tap root (main root)
- lateral roots
- **tap root system**

Typical Monocotyledon – Sweet vernal grass
Anthoxanthum odoratum

- stamens
- inflorescence
- spikelets
- leaf blade (lamina) with parallel venation
- sheathing leaf base completely surrounds stem
- node
- adventitious roots
- **fibrous root system**

Anthoxanthum odoratum grows to a height of 30–100 cm and flowers from April to July.

Epilobium angustifolium grows to a height of 1.5 m and flowers from June to September.

69

Branching

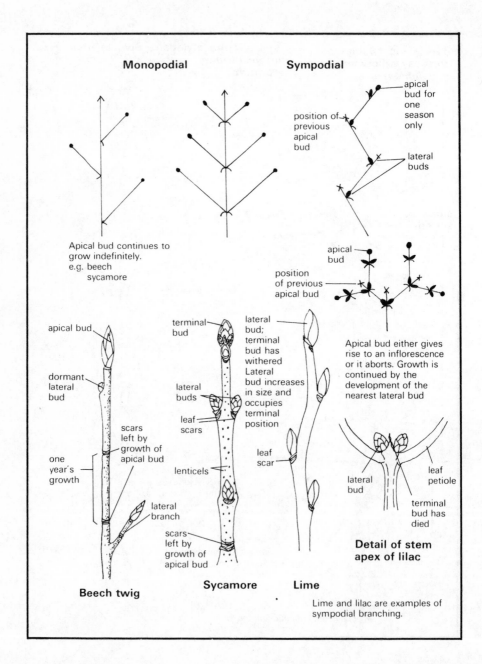

Monopodial

Apical bud continues to grow indefinitely.
e.g. beech
     sycamore

Sympodial

apical bud for one season only

position of previous apical bud

lateral buds

apical bud

position of previous apical bud

Apical bud either gives rise to an inflorescence or it aborts. Growth is continued by the development of the nearest lateral bud

apical bud

dormant lateral bud

scars left by growth of apical bud

one year's growth

lateral branch

Beech twig

terminal bud

lateral buds

leaf scars

lenticels

scars left by growth of apical bud

Sycamore

lateral bud; terminal bud has withered Lateral bud increases in size and occupies terminal position

leaf scar

Lime

lateral bud

leaf petiole

terminal bud has died

Detail of stem apex of lilac

Lime and lilac are examples of sympodial branching.

Leaf forms

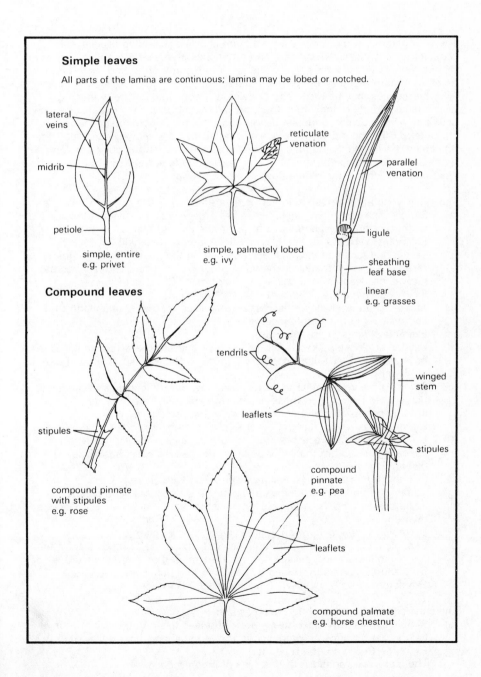

Simple leaves

All parts of the lamina are continuous; lamina may be lobed or notched.

lateral veins

midrib

petiole

simple, entire
e.g. privet

reticulate venation

simple, palmately lobed
e.g. ivy

parallel venation

ligule

sheathing leaf base

linear
e.g. grasses

Compound leaves

stipules

compound pinnate
with stipules
e.g. rose

tendrils

leaflets

winged stem

stipules

compound pinnate
e.g. pea

leaflets

compound palmate
e.g. horse chestnut

Inflorescences

The arrangement of the flowers on the stem is called the inflorescence. It is constant for any particular species, and often characteristic for families or genera. Sometimes the flowers are solitary, but it is more common to find them grouped together, thus forming a conspicuous mass which will attract pollinating insects.

Inflorescences may be racemose or cymose, depending on the branching which occurs. If the branching is monopodial, the inflorescence is a raceme and does not lose its ability to initiate flower buds. This type of inflorescence is characterized by the oldest flowers occurring at the base of the peduncle. If the branching is sympodial, the inflorescence is a cyme. When the apex produces a terminal flower, the ability to initiate more flower buds at this apex is lost. In this type of inflorescence, the oldest flowers are at the apex.

The chief types of racemose inflorescences are:
(a)   Simple racemes where each flower has a pedicel and individual flowers are arranged along the peduncle, usually in the axils of bracts.
       Examples: *Digitalis* sp. (foxglove), *Endymion* sp. (bluebell)
(b)   Compound racemes in which each branch on the peduncle is a simple raceme. This is sometimes known as a panicle, and can be described as a raceme of racemes. It is characteristic of many Graminae (grasses).
       Example: *Festuca* sp. (fescue)
(c)   Spikes where each flower is sessile (unstalked) and arranged along the peduncle as in the simple raceme.
       Example: *Plantago* sp. (plantain)
(d)   Catkins are loose spikes which hang down and are characteristic of wind-pollinated trees such as *Populus* sp. (poplar), *Quercus* sp. (oak) and *Betula* sp. (birch).
(e)   Corymbs are very similar to simple racemes, but the pedicels of the lower flowers are longer than those nearest the apex. The result of this arrangement is that all the flowers are at the same level.
       Examples: *Iberis* sp. (candytuft) and other members of the Cruciferae
(f)   Umbels are simple racemes in which the pedicels are of equal length and appear to arise from the same point on the peduncle, i.e. internodes are not distinguishable.
       Examples: *Primula* sp. (cowslip), *Prunus* sp. (wild cherry)
(g)   Compound umbels are characteristic of the Umbelliferae and can be described as umbels of umbels, where each branch of an umbel bears its own umbel.
       Examples: *Petroselinum* sp. (parsley), *Daucus* sp. (carrot)
(h)   Capitula (singular: capitulum) are characteristic of the Compositae. In this type of inflorescence, sessile flowers are arranged on the flattened apex of the peduncle, and surrounded by bracts forming an involucre.
       Examples: *Bellis* sp. (daisy), *Taraxacum* sp. (dandelion)

The chief types of cymose inflorescences are:
(a)   Simple cymes in which a new flower bud is initiated in the axil of a bract behind the first flower, produced in the terminal position on the axis.
       Example: *Geum* sp. (herb bennet)
       There are two variations of this type of inflorescence:

(b)   Scorpioid cymes in which successive axes are developed from bracts alternately on opposite sides to the preceding axis.
Example: *Myosotis* sp. (Forget me not)

(c)   Helicoid cymes in which successive axes are developed from bracts on the same side of the preceding axis.
Example: *Freesia*

(d)   Dichasial cymes occur after a flower bud has been produced in the terminal position on the axis, and development of two more flower buds occurs in the axils of bracts behind the terminal position.
Examples: *Silene* sp. (bladder campion), *Stellaria* sp. (stitchwort)

(e)   In addition to these recognizable types of inflorescence, there are mixtures of the racemose and cymose forms. Often there is a racemose type of development in the main axis and then the lateral branches develop in a cymose manner.
Example: *Aesculus* sp. (horse chestnut) which is a panicle of scorpoid cymes

Racemose inflorescences

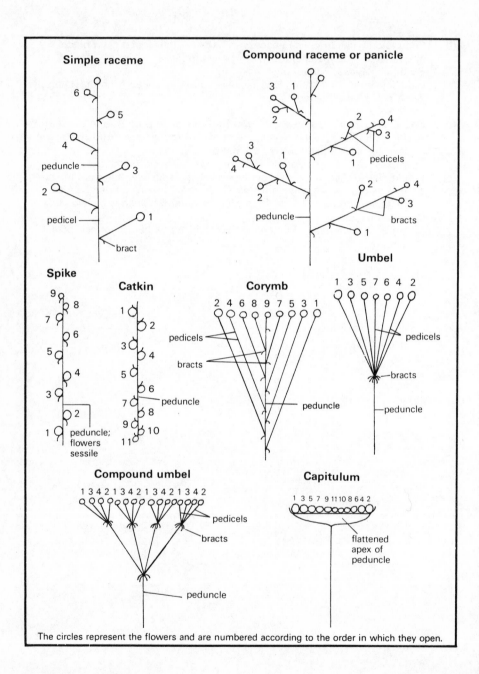

Simple raceme

Compound raceme or panicle

Spike

Catkin

Corymb

Umbel

Compound umbel

Capitulum

The circles represent the flowers and are numbered according to the order in which they open.

74

Cymose and mixed inflorescences

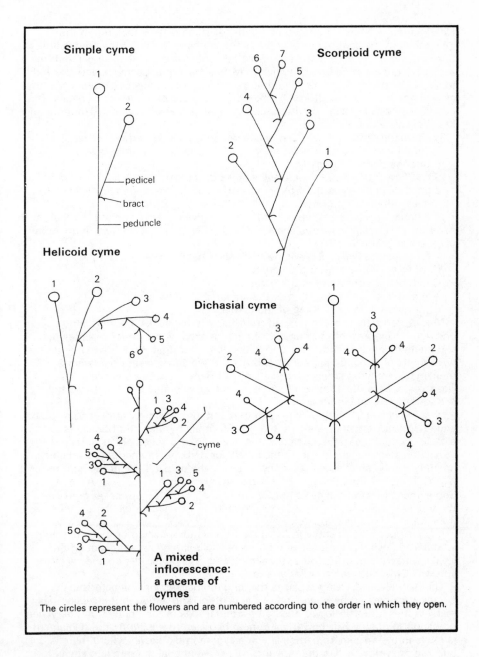

Simple cyme

Scorpioid cyme

Helicoid cyme

Dichasial cyme

A mixed inflorescence: a raceme of cymes

The circles represent the flowers and are numbered according to the order in which they open.

Floral morphology

Floral parts are usually arranged in whorls, with the carpels making up the gynaecium in the centre, surrounded by the stamens forming the androecium. Surrounding these essential parts of the flower is the perianth. The perianth may consist of one or more whorls which may be similar in appearance and referred to simply as the perianth; or often it is possible to distinguish an outer whorl, the calyx made up of sepals, from an inner whorl, the corolla made up of petals. The number of parts in each whorl tend to be equal, e.g. *Ranunculus* sp. (buttercup) has 5 sepals and 5 petals.

The structure of the flower can be shown conveniently in three ways:
- a floral formula – a shorthand method of indicating the numbers and arrangement of the parts
- a half-flower drawing – where the shapes of the parts are accurately drawn
- a floral diagram – a plan of the floral parts and their arrangement relative to each other.

In a floral formula, each type of floral part is given a letter and the number of parts for a particular flower is shown after it. The letters which denote the floral parts are as follows:
- P for perianth, where there is no distinction between sepals and petals
- K for calyx, consisting of the sepals
- C for corolla, consisting of the petals
- A for androecium, consisting of the stamens
- G for gynaecium, consisting of the carpels.

Much information can be derived from floral formulae, as it is normal to use one figure to represent the number of parts in each whorl. Perianth segments and stamens often occur in more than one whorl, so that P3 + 3 would indicate two whorls of perianth segments, and A3 + 3 would represent two whorls of stamens. In some families, particularly the Ranunculaceae and the Rosaceae, where the number of stamens and carpels may be quite large, then A ∞ and G ∞ signify numerous.

In some flowers, the floral parts are free, but in others they are joined or fused, and it is possible to indicate this in a floral formula by the use of brackets, e.g. C(5) would mean that the corolla is made up of 5 fused petals. In some cases, the stamens are joined to the corolla or perianth segments (epipetalous), and this can be shown by bracketing together the corolla (C) and the androecium (A), {C(5) A5} or the perianth (P) and the androecium (A), {P(3 + 3) A3 + 3}. Sometimes a different style of bracketing is used, where the bracket goes over the top, C̅(5) A5 or P̅(3 + 3) A3 + 3. The former convention will be used in the floral formulae mentioned later.

It is also possible to show in the floral formula the position of the gynaecium, whether the ovary is superior or inferior, by including a line representing the level of attachment of the other floral parts, e.g. G(2̲) indicates a superior ovary, whereas G(3̲) indicates an inferior one.

The half-flower drawing replaces the longitudinal section characteristic of botanical studies at the beginning of the century, and although the section has many virtues, the half-flower does give a better idea of the floral structure. When preparing to draw a half-flower, care must be taken to ensure that a median cut is made in the specimen, particularly in a zygomorphic flower where there is only one line of symmetry. On the drawing, it is conventional to represent any cut edge of petal, sepal or perianth segment by a double line. If the cut has passed between two parts, then a single line is drawn (see diagram on p. 78).

A floral diagram is similar to a map or plan, and shows the arrangement of the floral parts clearly. It can show the way in which the petals overlap, the placentation and whether the flower is zygomorphic, but it will not show the superior or inferior position of the ovary. The positions of the floral parts may change slightly as the flower opens fully, so an observation of flower buds or young flowers is sometimes necessary in order to draw an accurate floral diagram.

Floral morphology

Half-flower diagram of a generalized Angiosperm flower

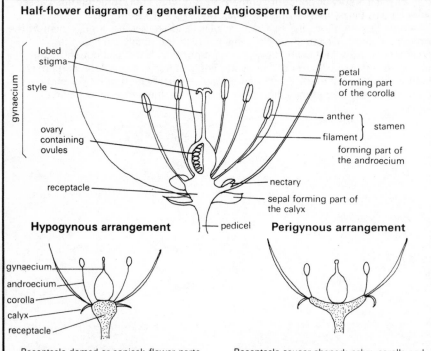

gynaecium

- lobed stigma
- style
- ovary containing ovules
- receptacle

- petal forming part of the corolla
- anther } stamen
- filament } forming part of the androecium
- nectary
- sepal forming part of the calyx
- pedicel

Hypogynous arrangement

- gynaecium
- androecium
- corolla
- calyx
- receptacle

Receptacle domed or conical; flower parts are inserted in the order: calyx, corolla, androecium, gynaecium. The gynaecium is above all other parts; ovary superior, e.g. buttercup, hyacinth.

Perigynous arrangement

Receptacle saucer-shaped; calyx, corolla and androecium develop on outer rim, gynaecium develops centrally on the receptacle, appearing at the same level as the other parts; ovary superior, e.g. *Potentilla* sp.

Epigynous arrangement

Receptacle flask-shaped; calyx, corolla and androecium inserted above the gynaecium; ovary inferior, e.g. apple.

Extreme perigynous arrangement

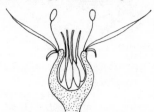

Receptacle deeply concave or urn-shaped; ovary has become more enclosed but has not been completely enveloped by the receptacle as in the epigynous arrangement; ovary superior, e.g. rose.

Development of the androecium

A stamen is a male sporangiophore and typically consists of a filament or stalk bearing four pollen sacs. The pollen sacs are usually in two pairs, side by side, at the apex of the filament, and form the anther. In the early stages of development, each pollen sac contains a central mass of microspore mother cells, surrounded by a nutritive layer called the tapetum. Each microspore mother cell undergoes meiosis to give a tetrad of haploid cells. Each of these cells develops into a pollen grain. A thick outer wall, or exine, and a thin inner wall, or intine, are secreted. The exine may be pitted and sculptured depending on the species.

Inside the pollen grain, the haploid nucleus undergoes mitosis to produce a generative nucleus and a pollen tube nucleus. The generative nucleus will eventually give rise to the 2 male gametes.

As the pollen grains mature, changes take place in the anthers. The cells of the tapetum shrink and degenerate, and fibrous layers develop beneath the epidermis. When the pollen is ripe, the outer layers of the anther dry out and tensions are set up in the lateral grooves. Eventually dehiscence occurs and the edges of the pollen sacs curl away, freely exposing the pollen grains.

Development of the androecium

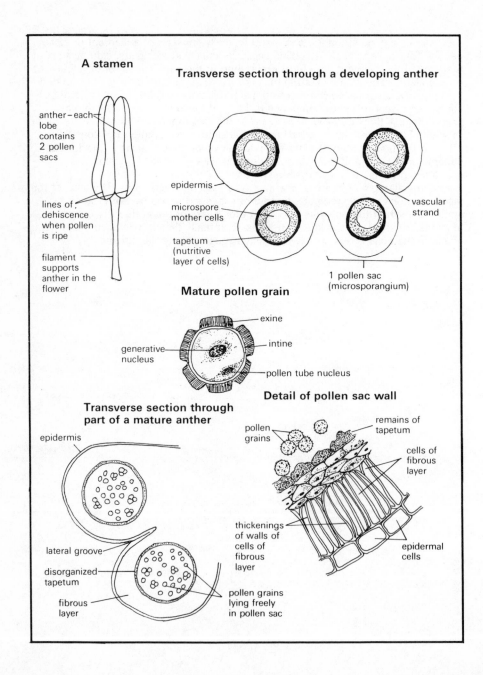

A stamen

anther – each lobe contains 2 pollen sacs

lines of dehiscence when pollen is ripe

filament supports anther in the flower

Transverse section through a developing anther

epidermis

microspore mother cells

tapetum (nutritive layer of cells)

vascular strand

1 pollen sac (microsporangium)

Mature pollen grain

exine

generative nucleus

intine

pollen tube nucleus

Detail of pollen sac wall

pollen grains

remains of tapetum

cells of fibrous layer

thickenings of walls of cells of fibrous layer

epidermal cells

pollen grains lying freely in pollen sac

Transverse section through part of a mature anther

epidermis

lateral groove

disorganized tapetum

fibrous layer

Development of the ovule

The female sporangiophore or megasporangiophore, is a closed structure called a carpel, typically with a receptive stigma on which the pollen germinates, borne on a style. Inside the carpel, one or more ovules may develop. Each ovule is an outgrowth of the placenta of the carpel and begins as a tiny lump of tissue called the nucellus. At the apex of the nucellus, a megaspore mother cell develops and undergoes meiosis to produce a linear tetrad of haploid megaspores. Normally three of the four megaspores abort, leaving the innermost one to continue its development. This single megaspore enlarges, gaining nutrition from the nucellus tissue. It undergoes three successive mitotic divisions and forms an embryo sac containing eight haploid nuclei. This embryo sac represents the female gametophyte, a very reduced structure in comparison with the gametophytes of the Bryophyta and the Pteridophyta.

As the nucellus enlarges, one or more integuments grow up from around its base. The integuments do not completely surround the nucellus, their development stops before they enclose the apex of the ovule, leaving a small pore or opening called the micropyle. At this stage the ovule is still very small and is attached to the placenta by a short stalk or funicle.

In the embryo sac, the eight nuclei, each associated with some cytoplasm, are arranged in a definite pattern. Three cells are situated at the micropylar end of the ovule and make up what is referred to as the egg apparatus. The central one of these cells is the female gamete or egg cell, and the surrounding cells are called synergids. At the opposite, or chalazal, end of the embryo sac is another group of three cells known as the antipodal cells. The remaining two nuclei, called the polar nuclei, migrate to the centre of the embryo sac and may remain separate at this stage, eventually fusing to give a central diploid fusion nucleus.

This type of development is only one of several kinds which are known to occur in the Angiospermae. Variations occur in the details as to which one of the megaspores survives and which abort, in the type and number of division, and in the number of nuclei eventually present in the embryo sac. Usually the egg nucleus remains haploid, but the chalazal polar nucleus may be diploid so that the central fusion nucleus becomes triploid.

Development of the ovule

Stages in the development of the ovule

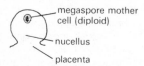

- megaspore mother cell (diploid)
- nucellus
- placenta

Megaspore mother cell undergoes meiosis and 4 haploid megaspores are produced.

- linear tetrad of potential megaspores
- integument begins to grow
- funicle

3 of the 4 megaspores abort; remaining megaspore grows bigger nourished by nucellus.

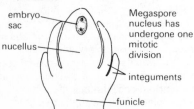

- embryo sac
- nucellus

Megaspore nucleus has undergone one mitotic division

- integuments
- funicle

Two more mitotic divisions occur in the embryo sac.

- micropyle
- embryo sac
- integuments almost cover ovule
- nucellus
- chalaza
- funicle

This erect form of ovule orientation is called orthotropous, e.g. *Polygonum* sp.

Enlarged view of mature embryo sac

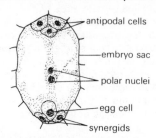

- antipodal cells
- embryo sac
- polar nuclei
- egg cell
- synergids

Anatropous ovule arrangement e.g. *Ranunculus* sp.

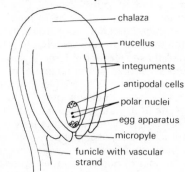

- chalaza
- nucellus
- integuments
- antipodal cells
- polar nuclei
- egg apparatus
- micropyle
- funicle with vascular strand

Campylotropous ovule arrangement e.g. *Malva* sp.

All ovules develop initially in an erect position on the placenta, but the anatropous and campylotropous forms arise due to more growth occurring on the side of the chalazal region than on the other.

The form of the gynaecium

The gynaecium consists of one or more carpels, which may be separate (apocarpous) or fused together (syncarpous) on the receptacle. The apocarpous condition is considered to be more primitive than the syncarpous, and is found in the Ranunculaceae and some members of the Rosaceae. Where the carpels are syncarpous, fusion has occurred laterally and one or more internal cavities, or loculi, are formed. The position in which the ovules develop depends on the way in which fusion has occurred, and may be described as marginal (parietal), axile or central. These terms are illustrated on page 84.

In apocarpous gynaecia, each carpel has its own style with the stigma at the apex. In syncarpous gynaecia, the styles may be separate, but are more usually united to form a single structure bearing lobed stigmas; the number of lobes corresponding to the number of carpels.

The form of the gynaecium

Apocarpous gynaecia e.g. *Pisum* sp. (pea)

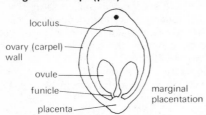

loculus
ovary (carpel) wall
ovule
funicle
placenta
marginal placentation

Transverse section through unicarpellary gynaecium of legume

e.g. *Caltha* sp. (marsh marigold)

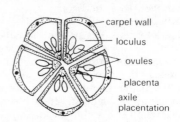

carpel wall
loculus
ovules
placenta
axile placentation

Transverse section through polycarpellary apocarpous gynaecium

Syncarpous gynaecia

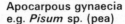

Diagrammatic longitudinal sections through tricarpellary gynaecia

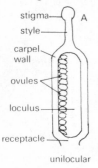

A
stigma
style
carpel wall
ovules
loculus
receptacle

unilocular

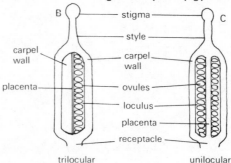

B
carpel wall
placenta

stigma
style
carpel wall
ovules
loculus
placenta
receptacle

C

trilocular

unilocular

Diagrammatic transverse sections through tricarpellary gynaecia

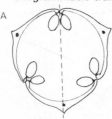

A

Marginal (parietal) placentation, e.g. some Iridaceae.

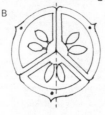

B

Axile placentation, e.g. Scrophulariaceae, Amaryllidaceae, Liliaceae.

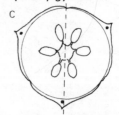

C

Central placentation; thought to be derived from B when inner carpel walls are lost leaving central column of placenta tissue on which ovules are borne.

Pollination

Pollination is necessary in order that the pollen grains, containing the male gametes, are brought into contact with the gynaecium so that fertilization can take place. In the Angiospermae, this means that the pollen grains must be transferred from the ripe anther to the receptive stigma. In some species, self-pollination occurs and pollen from the anthers of a flower need only be transferred to the stigma of the same flower, or another flower on the same plant. In a large number of species, however, cross-pollination is seen, i.e. pollen is transferred from the anthers of one flower to the stigma of a flower on another plant of the same species.

There are two main ways in which transfer of pollen occurs: by wind (anemophilous flowers) or by insects (entomophilous flowers). The two types of flower are well-adapted to their mode of pollination and it is easy to distinguish them by means of their differences in structure and arrangement of parts. These differences are summarized in the following table.

Differences between wind-pollinated and insect-pollinated flowers

Wind-pollinated	Insect-pollinated
Flowers usually inconspicuous; perianth segments, if present, small and green. Male flowers may be numerous and aggregated into pendulous catkins.	Flowers with large, often brightly coloured perianth segments, but sometimes small and aggregated to form compact inflorescences to attract pollinating insects.
Never scented.	May be scented to attract insects.
Nectaries not present; nectar not produced.	Nectaries often present; produce sugary nectar collected by insects for food.
Stamens either numerous, or with large anthers, or male flowers more numerous than female flowers. Large amounts of pollen produced.	Stamens fewer in number, usually with small anthers. Smaller amounts of pollen produced.
Filaments often long, extending outside perianth, if present; anthers loosely hinged, versatile; pollen easily dislodged and dispersed by air currents.	Filaments often short, fixed inside the perianth segments; anthers introrse or extrorse; situated to come into contact with visiting insects.
Pollen grains smooth-walled, small, dry and light; carried easily in air currents.	Pollen grains with sculptured walls, larger; stick to bodies of insects.
Stigmas branched and feathery, borne on long styles, often extending outside the perianth if present; airborne pollen easily trapped.	Stigmas capitate or lobed, usually borne on short styles, remaining enclosed by the perianth segments; stigma with grooved surface to which pollen grains adhere.
Anemophilous flowers are characteristic of trees, such as oak and elm where they are produced before the foliage leaves, or of herbaceous plants such as grasses, where they are on long peduncles above the surrounding leaves.	Entomophilous flowers often have highly specialized pollination mechanisms, adapted to the type of visiting insects, and there are also mechanisms which prevent self-pollination and promote cross-pollination.

Fertilization and seed formation

Successful pollination means that compatible pollen grains are deposited on the receptive surface of the stigma. If the pollen grains are incompatible, then they will not germinate, or if they do then growth soon ceases.

The stigma produces a sugary solution in which the pollen grains germinate. A fine germ tube, or pollen tube, pushes its way out through a pore in the exine and penetrates the stigma and style tissues, growing between the cells. The pollen tube nucleus is usually situated near the tip of the tube, with the two male gamete nuclei derived from the generative nucleus close behind.

The pollen tube grows down through the tissues of the style, secreting enzymes which digest some of the stylar tissue to provide nutrients for the growth. When the pollen tube reaches the loculus of the carpel, in most cases it will grow round towards the micropyle of the ovule and pass into the ovule, thus reaching the embryo sac. As the pollen tube penetrates the embryo sac, the tip opens and the two male gamete nuclei enter, the pollen tube nucleus having disintegrated earlier. The first male nucleus to enter the embryo sac fuses with the female or egg nucleus to form a zygote. The second male nucleus passes further into the embryo sac and fuses with the two polar nuclei, or with the central fusion nucleus if they have already fused, forming a polyploid primary endosperm nucleus. This double fertilization is a unique feature of the angiosperms. The other nuclei in the embryo sac degenerate, leaving the zygote and the endosperm nucleus to continue growth.

The diploid zygote nucleus begins to divide by mitosis and eventually an embryo, consisting of a plumule, a radicle and one or two cotyledons, is formed. The endosperm nucleus also divides mitotically, forming a large number of nuclei which lie freely in the cytoplasm of the embryo sac. The central part of the embryo sac becomes vacuolated and these nuclei become arranged peripherally and surrounded by cellulose cell walls. Division of cells continues and gradually an endosperm tissue is built up around the developing embryo. It is important to note that this tissue is not homologous with the endosperm in the gymnosperms, which is haploid and female prothallus tissue.

The endosperm cells become filled with stored food materials such as carbohydrates, lipids and proteins, which provide reserves for the developing embryo. The extent to which these reserves are drawn upon varies with the species. In endospermic seeds, there is quite a large amount of endosperm tissue remaining in the mature seed, whereas in the non-endospermic seeds most of the food reserves have been used up or transferred to the cotyledons which become swollen as a consequence.

As embryo development proceeds, changes take place in the integuments. If there was more than one integument initially, then fusion occurs to form a single layer, the cells of which become thickened with lignin and cutin. This layer, now called the testa, becomes very tough and impervious, and provides good protection for the dormant seed. During the later stages of development, the seed dries out, losing as much as 90 per cent by weight of its water content. It becomes a dormant structure, able to withstand adverse conditions, and in this state it is dispersed from the parent plant.

Changes also take place in the carpel wall (pericarp), and often the receptacle, to form the fruit from which the seeds may be eventually dispersed. In many cases the pericarp becomes dry and the seeds are either dispersed by wind, by sticking to the coats of animals or by explosive mechanisms. In other cases, the pericarp, together with the receptacle, becomes fleshy and palatable to animals that carry the fruit some distance away from the parent plant before eating it.

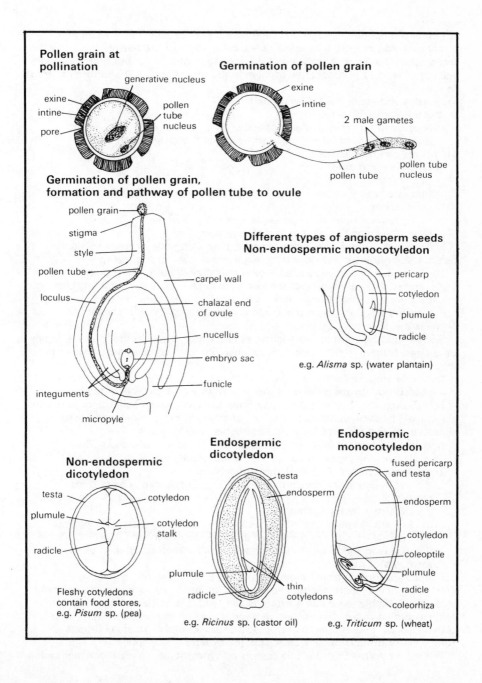

Pollen grain at pollination

- exine
- intine
- pore
- generative nucleus
- pollen tube nucleus

Germination of pollen grain

- exine
- intine
- 2 male gametes
- pollen tube
- pollen tube nucleus

Germination of pollen grain, formation and pathway of pollen tube to ovule

- pollen grain
- stigma
- style
- pollen tube
- loculus
- integuments
- micropyle
- carpel wall
- chalazal end of ovule
- nucellus
- embryo sac
- funicle

**Different types of angiosperm seeds
Non-endospermic monocotyledon**

- pericarp
- cotyledon
- plumule
- radicle

e.g. *Alisma* sp. (water plantain)

Non-endospermic dicotyledon

- testa
- plumule
- radicle
- cotyledon
- cotyledon stalk

Fleshy cotyledons contain food stores, e.g. *Pisum* sp. (pea)

Endospermic dicotyledon

- testa
- endosperm
- plumule
- radicle
- thin cotyledons

e.g. *Ricinus* sp. (castor oil)

Endospermic monocotyledon

- fused pericarp and testa
- endosperm
- cotyledon
- coleoptile
- plumule
- radicle
- coleorhiza

e.g. *Triticum* sp. (wheat)

Fruits

A fruit is the ripened gynaecium of a flower and is usually only formed after fertilization has occurred. In some cases, other parts of the flower, especially the receptacle, may become involved in the formation of a 'fruit', but these are more accurately termed pseudocarps, or false fruits, to distinguish them from the true fruits.

True fruits consist of the following parts:

    the ovary wall, or pericarp;

    one or more fertilized ovules, or seeds;

    remains or scars of other floral parts such as the style or stigma.

True fruits can be classified as follows.

    Pericarp dry and hard:

    – dehiscent

       fruit opens to disperse its seeds, e.g. follicle, legume, capsule, schizocarp

    – indehiscent

       fruit does not open to disperse its seeds, e.g. achene

    Pericarp fleshy and succulent, e.g. drupe, berry

Follicles are formed from single carpels, containing one or more seeds, and dehisce along the ventral sides only, e.g. *Delphinium* sp., *Caltha* sp.

Legumes are similar to follicles except that they dehisce along both dorsal and ventral sides. They are typical of the family Leguminosae, e.g. *Pisum* sp., *Ulex* sp.

Capsules are formed from syncarpous gynaecia and there are slits, pores or teeth formed through which the seeds are dispersed, e.g. *Antirrhinum* sp., *Digitalis* sp., *Iris* sp.

Siliquas and Siliculas are special types of capsule, typically found in the family Cruciferae, formed from bicarpellary ovaries where a false septum divides the loculus into two compartments.

Carcerula are also forms of capsule found in the Labiatae, where a number of one-seeded nutlets are formed by the growth of a dividing wall in the ovary.

Schizocarps are formed from syncarpous gynaecia which split up, or in which the carpels become separated from one another, forming one-seeded units, e.g. *Alcea* sp. (hollyhock) and the double samaras of *Acer* sp.

Achenes are formed from single carpels which contain one seed. They frequently occur in heads or clusters on a receptacle, e.g. *Ranunculus* sp.

    There are a number of special types of achene:

       the nut in which the pericarp is woody, e.g. *Corylus* sp., *Quercus* sp.;

       the caryopsis in which the pericarp and testa are fused, e.g. *Avena* sp. and typical of the family Graminae;

       the samara in which part of the pericarp forms a wing, e.g. *Fraxinus* sp.;

       the cypsela which is derived from a bicarpellary ovary and in which part of the wall is formed from the receptacle, e.g. *Helianthus* sp. and other members of the family Compositae.

Drupes are formed from single carpels or from syncarpous gynaecia, containing one or more seeds. The pericarp is divided into three parts; an inner endocarp, which is hard and surrounds the seed, a soft, fleshy mesocarp in the middle, and an outer, thin epicarp forming a skin on the outside, e.g. *Prunus* sp., *Sambucus* sp. Sometimes the receptacle bears a number of small drupes (drupelets or drupels) as in *Rubus* spp. (blackberry and raspberry).

Berries are formed from single carpels or syncarpous gynaecia, containing one or more seeds as in drupes, but each seed is surrounded by its own testa. The inner part of the pericarp does not become hard as in the drupes, e.g. *Cucurbita* sp. (marrow), *Vitis* sp. (vine).

The most common types of false fruits involve the receptacle, which often becomes large and fleshy. The strawberry (*Fragaria* sp.) is a typical example; the true fruits are achenes borne on the greatly enlarged, succulent receptacle. In the rose-hip (*Rosa* sp.), derived from an extremely perigynous flower, the receptacle forms an urn-shaped container for the achenes. In the pomes of apples and pears (*Malus* sp. and *Pyrus* sp.), the core is formed by the inferior syncarpous gynaecium, the inner part of the carpel wall becoming tough. The flesh is formed partly from the outer wall of the carpel and partly from the wide, succulent receptacle fused to it.

Dispersal

After successful fertilization and seed formation, dispersal of the fruits and seeds is important in ensuring that the species become as widespread as possible. Efficient dispersal avoids overcrowding and reduces competition. Usually large numbers of seeds are produced as many will not reach suitable situations for germination and growth. In order to bring about wide dispersal, fruits and seeds have evolved many adaptations in structure. Some of these adaptations have already been mentioned, and relevant dispersal mechanisms will be explained for the examples chosen to illustrate the angiosperm families described later.

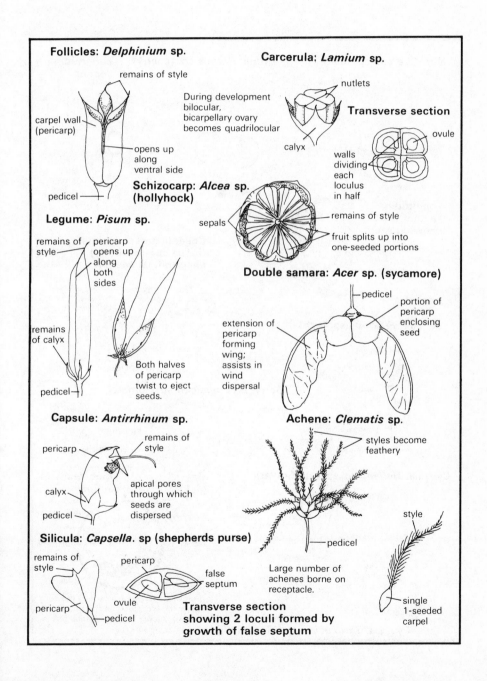

Follicles: *Delphinium* sp.

remains of style

carpel wall (pericarp)

opens up along ventral side

pedicel

Carcerula: *Lamium* sp.

nutlets

Transverse section

During development bilocular, bicarpellary ovary becomes quadrilocular

calyx

ovule

walls dividing each loculus in half

Schizocarp: *Alcea* sp. (hollyhock)

sepals

remains of style

fruit splits up into one-seeded portions

Legume: *Pisum* sp.

remains of style

pericarp opens up along both sides

remains of calyx

pedicel

Both halves of pericarp twist to eject seeds.

Double samara: *Acer* sp. (sycamore)

pedicel

portion of pericarp enclosing seed

extension of pericarp forming wing; assists in wind dispersal

Capsule: *Antirrhinum* sp.

pericarp

remains of style

calyx

apical pores through which seeds are dispersed

pedicel

Achene: *Clematis* sp.

styles become feathery

style

pedicel

Large number of achenes borne on receptacle.

single 1-seeded carpel

Silicula: *Capsella*. sp (shepherds purse)

remains of style

pericarp

false septum

ovule

pericarp

pedicel

Transverse section showing 2 loculi formed by growth of false septum

Fruits

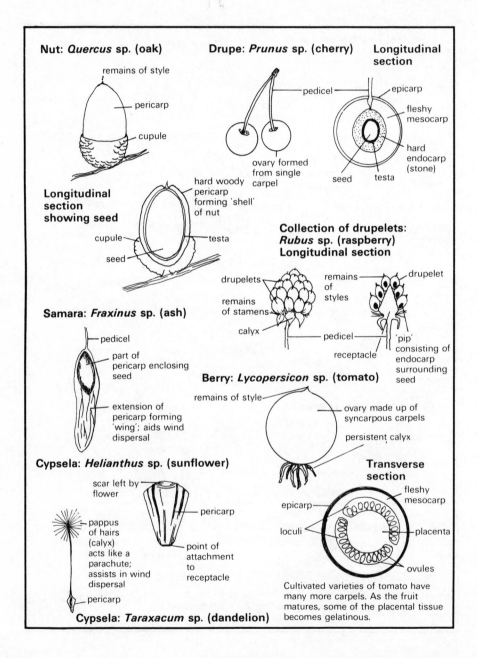

Nut: *Quercus* sp. (oak)

- remains of style
- pericarp
- cupule

Longitudinal section showing seed

- cupule
- seed
- hard woody pericarp forming 'shell' of nut
- testa

Samara: *Fraxinus* sp. (ash)

- pedicel
- part of pericarp enclosing seed
- extension of pericarp forming 'wing'; aids wind dispersal

Cypsela: *Helianthus* sp. (sunflower)

- scar left by flower
- pappus of hairs (calyx) acts like a parachute; assists in wind dispersal
- pericarp
- pericarp
- point of attachment to receptacle

Cypsela: *Taraxacum* sp. (dandelion)

Drupe: *Prunus* sp. (cherry)

- pedicel
- ovary formed from single carpel

Longitudinal section

- epicarp
- fleshy mesocarp
- hard endocarp (stone)
- seed
- testa

Collection of drupelets: *Rubus* sp. (raspberry) Longitudinal section

- drupelets
- remains of stamens
- calyx
- remains of styles
- drupelet
- pedicel
- receptacle
- 'pip' consisting of endocarp surrounding seed

Berry: *Lycopersicon* sp. (tomato)

- remains of style
- ovary made up of syncarpous carpels
- persistent calyx

Transverse section

- epicarp
- loculi
- fleshy mesocarp
- placenta
- ovules

Cultivated varieties of tomato have many more carpels. As the fruit matures, some of the placental tissue becomes gelatinous.

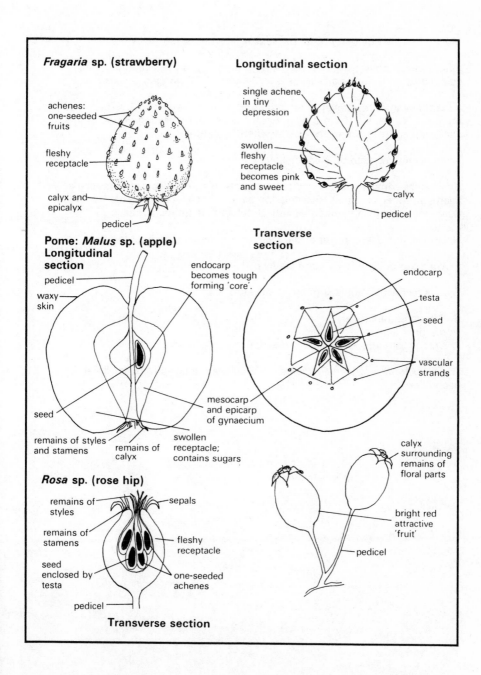

Fragaria sp. (strawberry)

Longitudinal section

achenes:
one-seeded
fruits

single achene
in tiny
depression

fleshy
receptacle

swollen
fleshy
receptacle
becomes pink
and sweet

calyx and
epicalyx

calyx

pedicel

pedicel

Pome: Malus sp. (apple)
Longitudinal section

Transverse section

pedicel

endocarp
becomes tough
forming 'core'.

endocarp

waxy
skin

testa

seed

vascular
strands

seed

mesocarp
and epicarp
of gynaecium

remains of styles
and stamens

remains of
calyx

swollen
receptacle;
contains sugars

calyx
surrounding
remains of
floral parts

Rosa sp. (rose hip)

remains of
styles

sepals

remains of
stamens

fleshy
receptacle

bright red
attractive
'fruit'

seed
enclosed by
testa

one-seeded
achenes

pedicel

pedicel

Transverse section

Sub-class Monocotyledones
Family: Iridaceae

Mostly herbaceous perennials, possessing rhizomes, corms or bulbs.

Inflorescence solitary or cymose; flowers hermaphrodite, actinomorphic, epigynous, with floral parts in 3s or multiples of 3 arranged cyclically.

Perianth segments consist of an inner and an outer whorl of 3, usually petaloid, united at the base to form a tube. Spathe present, consisting of two bracts enclosing and protecting the flower when immature.

Androecium consists of 3 epiphyllous stamens.

Gynaecium consists of 3 syncarpous carpels; either unilocular or tri-locular, with a large number of ovules. Placentation parietal if one loculus, axile if 3 loculi. Ovary inferior. Style sometimes with 3 petaloid stigmatic surfaces.

Nectaries, if present, occur at the base of the perianth segments.

Flowers are pollinated by large insects.

Fruit is a capsule, dehiscing into 3 valves.

British genera include:
  Iris (iris, flag)
  Crocus (crocus)

Freesia, *Montbretia* and *Gladiolus* are cultivated species used for floral decoration.

Crocus
purpureus
Crocus

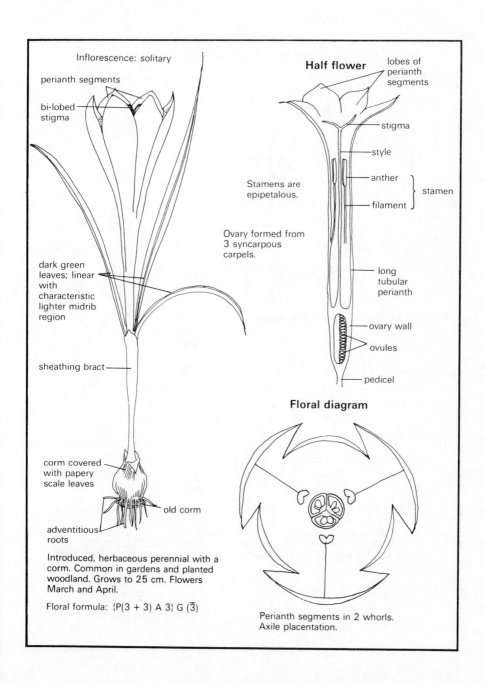

Inflorescence: solitary

perianth segments

bi-lobed stigma

dark green leaves; linear with characteristic lighter midrib region

sheathing bract

corm covered with papery scale leaves

old corm

adventitious roots

Introduced, herbaceous perennial with a corm. Common in gardens and planted woodland. Grows to 25 cm. Flowers March and April.

Floral formula: ⟨P(3 + 3) A 3⟩ G ($\overline{3}$)

Half flower

lobes of perianth segments

stigma

style

anther

filament

stamen

Stamens are epipetalous.

Ovary formed from 3 syncarpous carpels.

long tubular perianth

ovary wall

ovules

pedicel

Floral diagram

Perianth segments in 2 whorls. Axile placentation.

Crocus
purpureus
Crocus

DIVISION SPERMATOPHYTA

CLASS ANGIOSPERMAE

SUB-CLASS MONOCOTYLEDONES

FAMILY IRIDACEAE

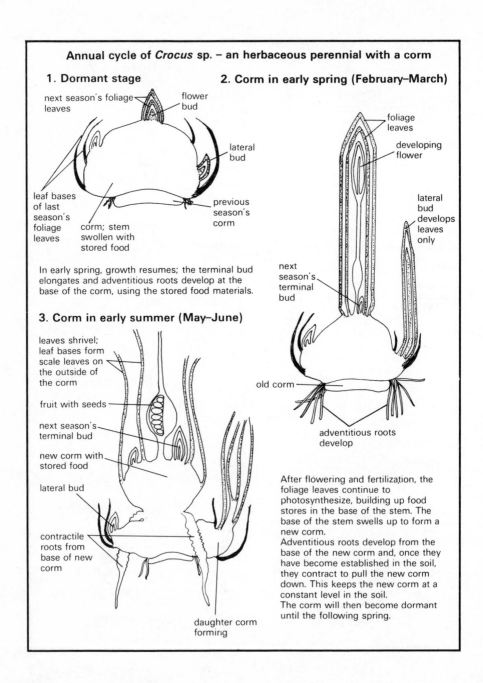

Annual cycle of *Crocus* sp. – an herbaceous perennial with a corm

1. Dormant stage

next season's foliage leaves

flower bud

lateral bud

leaf bases of last season's foliage leaves

corm; stem swollen with stored food

previous season's corm

In early spring, growth resumes; the terminal bud elongates and adventitious roots develop at the base of the corm, using the stored food materials.

2. Corm in early spring (February–March)

foliage leaves

developing flower

lateral bud develops leaves only

next season's terminal bud

old corm

adventitious roots develop

3. Corm in early summer (May–June)

leaves shrivel; leaf bases form scale leaves on the outside of the corm

fruit with seeds

next season's terminal bud

new corm with stored food

lateral bud

contractile roots from base of new corm

daughter corm forming

After flowering and fertilization, the foliage leaves continue to photosynthesize, building up food stores in the base of the stem. The base of the stem swells up to form a new corm.
Adventitious roots develop from the base of the new corm and, once they have become established in the soil, they contract to pull the new corm down. This keeps the new corm at a constant level in the soil.
The corm will then become dormant until the following spring.

Family: Liliaceae

Mostly herbaceous, possessing rhizomes, corms or bulbs.

Inflorescence solitary or racemose; flowers hermaphrodite, actinomorphic, hypogynous, with floral parts in 3s or mulitples of 3.

Perianth segments petaloid, arranged in two similar whorls of 3; may be free or joined at the base.

Androecium consists of two whorls of stamens, inserted freely opposite the perianth segments, or inserted on them.

Gynaecium consists of 3 syncarpous carpels with 3 loculi containing a large number of ovules. Placentation axile. Ovary superior, with a single style, if present. Stigmatic surface 3-lobed.

Nectaries, if present, situated at the top of the ovary.

Flowers are either insect-pollinated or self-pollinated.

Fruit may be a berry or a capsule; seeds endospermic.

British genera include:
    Endymion (bluebell)
    Allium (garlic, ramsons)
    Polygonatum (solomon's seal)
    Convallaria (lily of the valley)
    Ruscus (butcher's broom)

Species of *Lilium* and *Tulipa* are cultivated for floral decoration, and *Allium* spp. such as leeks and onions are grown for food.

Endymion non-scriptus
Bluebell or wild hyacinth

DIVISION SPERMATOPHYTA

CLASS ANGIOSPERMAE

SUB-CLASS MONOCOTYLEDONES

FAMILY LILIACEAE

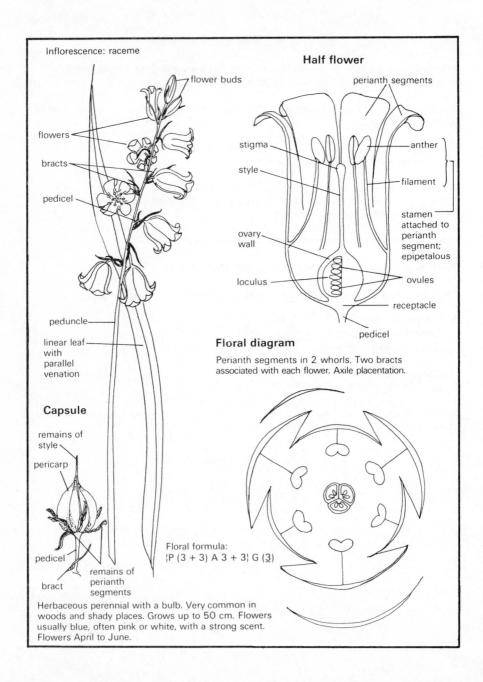

Inflorescence: raceme

flower buds

flowers

bracts

pedicel

peduncle

linear leaf with parallel venation

Capsule

remains of style

pericarp

pedicel

bract

remains of perianth segments

Half flower

perianth segments

stigma

style

anther

filament

stamen attached to perianth segment; epipetalous

ovary wall

loculus

ovules

receptacle

pedicel

Floral diagram

Perianth segments in 2 whorls. Two bracts associated with each flower. Axile placentation.

Floral formula:
¦P (3 + 3) A 3 + 3¦ G (3̱)

Herbaceous perennial with a bulb. Very common in woods and shady places. Grows up to 50 cm. Flowers usually blue, often pink or white, with a strong scent. Flowers April to June.

Family: Amaryllidaceae

Herbaceous perennials possessing bulbs.

Inflorescence solitary or umbellate; flowers hermaphrodite, actinomorphic, epigynous, with floral parts in 3s or multiples of 3.

Perianth segments petaloid, arranged in two whorls of 3; may be free or joined. Corona sometimes present. Thin spathe present, enclosing flowers when immature.

Androecium consists of 2 whorls of stamens, inserted freely opposite the perianth segments or inserted on them.

Gynaecium consists of 3 syncarpous carpels, with 3 loculi and numerous ovules. Placentation axile. Ovary inferior. Simple style with capitate or 3-lobed stigma.

Flowers either self-pollinated or insect-pollinated.

Fruit is a capsule; seeds endospermic.

British genera include:
  Galanthus (snowdrop)
  Leucojum (snowflake)
  Narcissus (daffodil)

This family differs from the Liliaceae in having an inferior ovary, and from the Iridaceae in having 2 whorls of 3 stamens.

Narcissus pseudonarcissus
Daffodil

DIVISION SPERMATOPHYTA

CLASS ANGIOSPERMAE

SUB-CLASS MONOCOTYLEDONES

FAMILY AMARYLLIDACEAE

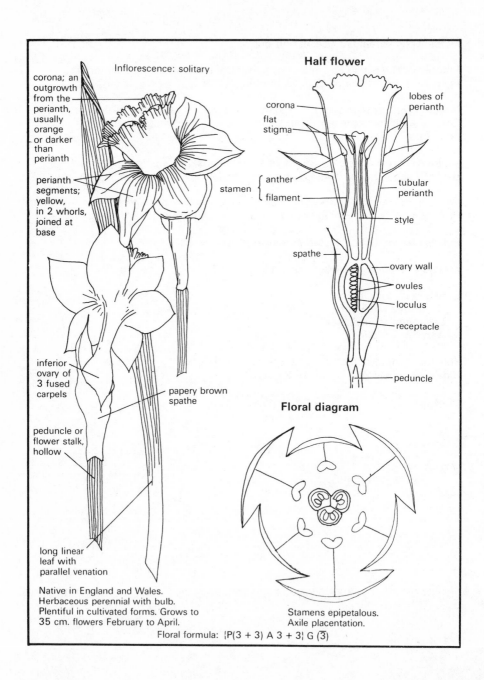

Inflorescence: solitary

corona; an outgrowth from the perianth, usually orange or darker than perianth

perianth segments; yellow, in 2 whorls, joined at base

inferior ovary of 3 fused carpels

papery brown spathe

peduncle or flower stalk, hollow

long linear leaf with parallel venation

Half flower

corona

flat stigma

stamen { anther / filament

lobes of perianth

tubular perianth

style

spathe

ovary wall

ovules

loculus

receptacle

peduncle

Floral diagram

Native in England and Wales. Herbaceous perennial with bulb. Plentiful in cultivated forms. Grows to 35 cm. flowers February to April.

Stamens epipetalous. Axile placentation.

Floral formula: ⎰P(3 + 3) A 3 + 3⎱ G (3̄)

102

Narcissus pseudonarcissus
Daffodil

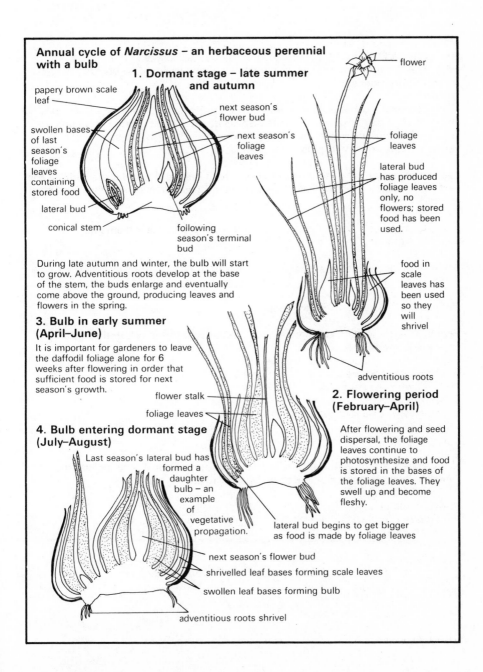

Annual cycle of *Narcissus* – an herbaceous perennial with a bulb

1. Dormant stage – late summer and autumn

flower

papery brown scale leaf

swollen bases of last season's foliage leaves containing stored food

lateral bud

conical stem

next season's flower bud

next season's foliage leaves

following season's terminal bud

foliage leaves

lateral bud has produced foliage leaves only, no flowers; stored food has been used.

During late autumn and winter, the bulb will start to grow. Adventitious roots develop at the base of the stem, the buds enlarge and eventually come above the ground, producing leaves and flowers in the spring.

food in scale leaves has been used so they will shrivel

3. Bulb in early summer (April–June)

It is important for gardeners to leave the daffodil foliage alone for 6 weeks after flowering in order that sufficient food is stored for next season's growth.

flower stalk

foliage leaves

adventitious roots

2. Flowering period (February–April)

After flowering and seed dispersal, the foliage leaves continue to photosynthesize and food is stored in the bases of the foliage leaves. They swell up and become fleshy.

4. Bulb entering dormant stage (July–August)

Last season's lateral bud has formed a daughter bulb – an example of vegetative propagation.

lateral bud begins to get bigger as food is made by foliage leaves

next season's flower bud

shrivelled leaf bases forming scale leaves

swollen leaf bases forming bulb

adventitious roots shrivel

103

Family: *Graminae*

Mostly herbaceous; annual or perennial; possessing linear leaves with sheathing leaf bases and ligules. Stems are solid at the nodes, hollow at internodes; cylindrical or flattened. Intercalary meristems above internodes; branching frequent from nodes.

Unit of inflorescence a spikelet consisting of 1–5 sessile florets on a short axis, enclosed by a pair of glumes (bracts). Spikelets arranged to form compound spikes (*Triticum* sp.), racemes (*Festuca* sp.) or panicles (*Avena* sp.). Florets hermaphrodite, actinomorphic, hypogynous.

Each floret has at the base a pair of small green bracts; the lower bract called the lemma, the upper bract called the palea. Within these bracts there may be a pair of lodicules.

Androecium usually consists of 3 stamens with long delicate filaments and versatile anthers.

Gynaecium consists of 3 syncarpous carpels with one loculus, enclosing a single ovule. Ovary superior. 2, rarely 3, short styles bearing feathery stigmas.

Flowers wind-pollinated; protogynous.

Fruit is an achene, with the pericarp and testa fused (often referred to as a caryopsis); seeds endospermic.

British genera include:
  Agropyron (couch grass)
  Bromus (brome)
  Dactylis (cock's foot grass)
  Festuca (fescue)
  Lolium (rye grass)

Many genera are of importance to humans as food plants. These include the cereals which form the staple diet of most populations, e.g.
  Triticum (wheat)
  Avena (oats)
  Hordeum (barley)
  Secale (rye)
  Zea (maize)
  Oryza (rice
  and *Saccharum* (sugar cane) from which sucrose is extracted.

Poa annua
Annual meadow grass

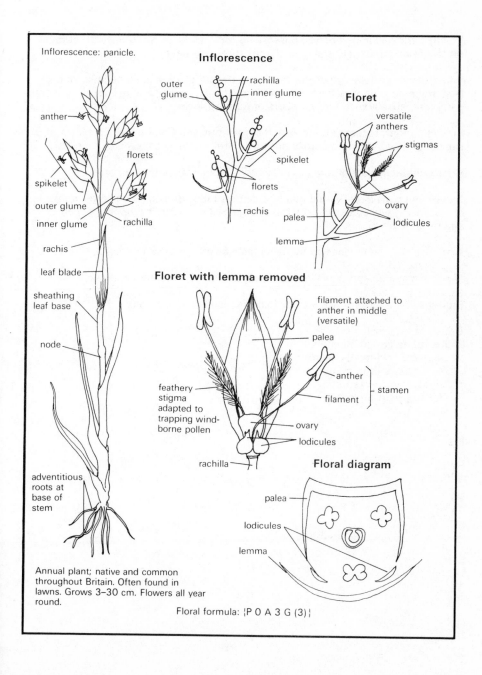

Inflorescence: panicle.

Inflorescence

outer glume
rachilla
inner glume

anther

florets

spikelet

outer glume

inner glume

rachilla

rachis

leaf blade

sheathing leaf base

node

spikelet

florets

rachis

Floret

versatile anthers

stigmas

ovary

palea

lodicules

lemma

Floret with lemma removed

filament attached to anther in middle (versatile)

palea

anther

filament

stamen

feathery stigma adapted to trapping wind-borne pollen

ovary

lodicules

rachilla

adventitious roots at base of stem

Annual plant; native and common throughout Britain. Often found in lawns. Grows 3–30 cm. Flowers all year round.

Floral diagram

palea

lodicules

lemma

Floral formula: ¦P 0 A 3 G (3)¦

105

Sub-class Dicotyledones
Family: Ranunculaceae

Mostly herbaceous perennials, possessing root stocks or rhizomes with fibrous roots. Leaves often palmately lobed with sheathing bases.

Inflorescence usually cymose or racemose (*Delphinium* sp.), sometimes solitary (*Anemone* sp.); flowers hermaphrodite, mostly actinomorphic, hypogynous; floral parts free, spirally inserted; numbers of parts variable.

Perianth undifferentiated or calyx and corolla present. Calyx usually consists of 5 sepals; petaloid when no corolla present. Corolla usually consists of 5 petals.

Androecium consists of numerous stamens, spirally inserted, extrorse.

Gynaecium consists of numerous free carpels (apocarpous). Each carpel contains one to many ovules. Basal or marginal placentation. Ovary superior; style and stigma simple.

Nectaries may be present on the receptacle or on the calyx or corolla.

Flowers may be wind-pollinated or insect-pollinated; often protandrous.

Fruit may be a head of achenes or follicles; seeds endospermic.

British genera include:
   Anemone (wood anemone)
   Clematis (old man's beard)
   Ranunculus (buttercup)
   Delphinium (larkspur)
   Helleborus (Christmas rose)

DIVISION SPERMATOPHYTA
CLASS ANGIOSPERMAE
SUB-CLASS DICOTYLEDONES
FAMILY RANUNCULACEAE

Ranunculus ficaria
Lesser celandine

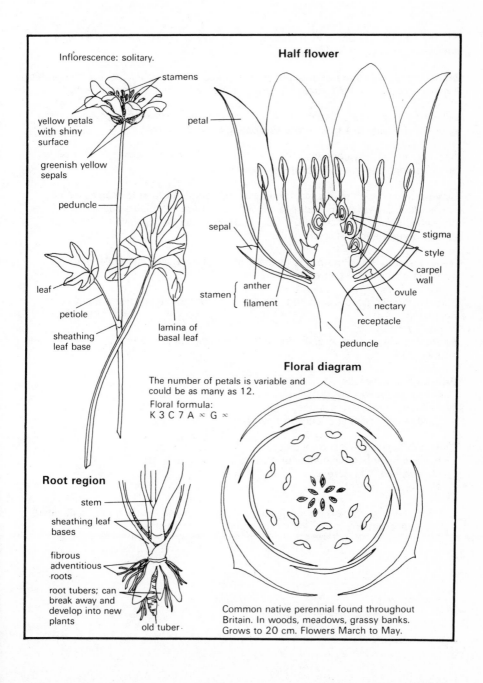

Inflorescence: solitary.

stamens

yellow petals
with shiny
surface

greenish yellow
sepals

peduncle

leaf

petiole

sheathing
leaf base

lamina of
basal leaf

Half flower

petal

sepal

stamen { anther
         filament

stigma
style
carpel
wall
ovule
nectary
receptacle
peduncle

Floral diagram

The number of petals is variable and
could be as many as 12.
Floral formula:
K 3 C 7 A ∞ G ∞

Root region

stem

sheathing leaf
bases

fibrous
adventitious
roots

root tubers; can
break away and
develop into new
plants

old tuber

Common native perennial found throughout
Britain. In woods, meadows, grassy banks.
Grows to 20 cm. Flowers March to May.

Family: Cruciferae

Mostly herbaceous annuals, biennials or perennials, possessing spirally arranged simple or pinnately lobed leaves.

Inflorescence a raceme or corymb; flowers usually hermaphrodite, actinomorphic, hypogynous; floral parts except carpels are freely inserted on the receptacle.

Calyx of 4 free sepals, usually in two pairs; the inner pair with nectar-collecting pouches. Corolla of 4 free petals alternating with the sepals.

Androecium usually consists of 6 stamens, 2 with short filaments and 4 with long filaments; rarely 4 or fewer stamens present.

Gynaecium consists of 2 syncarpous carpels containing one to many ovules; single loculus usually divided into two by the growth of a false septum from the placentae. Marginal placentation. Ovary superior; single style with 2-lobed or flattened stigmatic surface.

Nectaries situated at the bases of the 2 short stamens; nectar collects in the pouches of the inner pair of sepals.

Flowers mostly insect-pollinated and protandrous. Some species are self-pollinated.

Fruit a silicula or siliqua, opening by 2 valves from below; seeds non-endospermic.

British genera include:
    Barbarea (yellow rocket)
    Capsella (shepherd's purse)
    Cardamine (milkmaid, cuckoo flower)

Genera cultivated for food include:
    Brassica (cabbage, cauliflower, turnip, brussels sprouts)
    Nasturtium (watercress)
    Raphanus (radish)

Cheiranthus (wallflower), *Iberis* (candytuft) and *Lunaria* (honesty) are cultivated for floral decoration.

DIVISION SPERMATOPHYTA
CLASS ANGIOSPERMAE
SUB-CLASS DICOTYLEDONES
FAMILY CRUCIFERAE

Cheiranthus cheiri
Wallflower

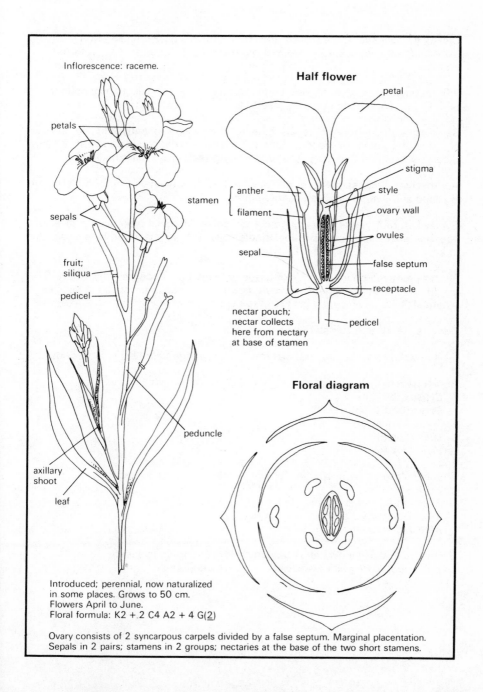

Inflorescence: raceme.

petals

sepals

fruit; siliqua

pedicel

stamen { anther, filament }

peduncle

axillary shoot

leaf

Half flower

petal

stigma

style

ovary wall

ovules

false septum

receptacle

anther

filament

sepal

nectar pouch; nectar collects here from nectary at base of stamen

pedicel

Floral diagram

Introduced; perennial, now naturalized in some places. Grows to 50 cm.
Flowers April to June.
Floral formula: K2 + 2 C4 A2 + 4 G($\underline{2}$)

Ovary consists of 2 syncarpous carpels divided by a false septum. Marginal placentation.
Sepals in 2 pairs; stamens in 2 groups; nectaries at the base of the two short stamens.

Family: *Leguminosae*

Mostly herbaceous, some shrubs and trees; annuals or perennials, with compound pinnate or trifoliate leaves often ending in tendrils; roots with root nodules.

Inflorescence racemose; flowers hermaphrodite, zygomorphic, hypogynous to perigynous; floral parts cyclically inserted.

Calyx usually consists of 5 sepals joined at the base, frequently 2-lipped. Corolla consists of 5 petals arranged as follows: 1 large adaxial petal (standard), 2 lateral petals (wings) and 2 lower petals joined along their lower margins (keel).

Androecium of 10 stamens, either all joined by their filaments to form a tube, or 9 joined and 1 free; rarely all stamens free.

Gynaecium of 1 carpel containing several ovules. Marginal placentation. Ovary superior, enclosed in stamen tube; single style with stigma, often long and coiled in keel.

Nectar produced on inner sides of bases of filaments in species which have 9 joined and 1 free stamens (*Lathyrus odoratus*). No nectar produced in species where all 10 stamens are joined (*Cytisus* spp.).

Flowers are insect-pollinated, usually by bees.

Fruit a legume; seeds non-endospermic, usually large with fleshy cotyledons.

British genera include:
    Cytisus (broom)
    Lotus (bird's foot trefoil)
    Trifolium (clover)
    Ulex (gorse)
    Vicia (vetch)

Genera cultivated by people for food and fodder for animals include:
    Pisum (pea)
    Phaseolus (runner bean)
    Vicia faba (broad bean)
    Medicago (lucerne)

Genera cultivated by people as garden trees and shrubs include:
    Laburnum, *Lathyrus* (sweet pea) and *Lupinus* (lupin)

DIVISION SPERMATOPHYTA
CLASS ANGIOSPERMAE
SUB-CLASS DICOTYLEDONES
FAMILY LEGUMINOSAE

Cytisus scoparius
Broom

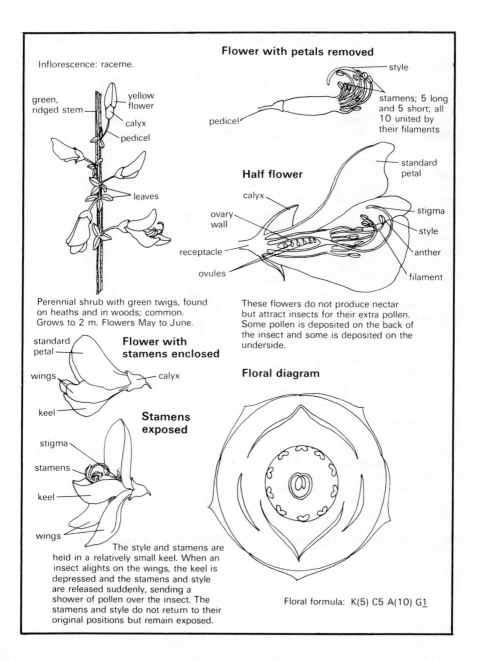

Inflorescence: raceme.

green, ridged stem

yellow flower

calyx

pedicel

leaves

Flower with petals removed

style

stamens; 5 long and 5 short; all 10 united by their filaments

pedicel

Half flower

standard petal

calyx

stigma

style

anther

filament

ovary wall

receptacle

ovules

Perennial shrub with green twigs, found on heaths and in woods; common. Grows to 2 m. Flowers May to June.

These flowers do not produce nectar but attract insects for their extra pollen. Some pollen is deposited on the back of the insect and some is deposited on the underside.

Flower with stamens enclosed

standard petal

wings

keel

calyx

Floral diagram

Stamens exposed

stigma

stamens

keel

wings

The style and stamens are held in a relatively small keel. When an insect alights on the wings, the keel is depressed and the stamens and style are released suddenly, sending a shower of pollen over the insect. The stamens and style do not return to their original positions but remain exposed.

Floral formula: K(5) C5 A(10) G<u>1</u>

Family: Rosaceae

Perennial herbs, shrubs or trees, usually with spirally arranged stipulate leaves.

Inflorescence cymose or racemose; flowers mostly hermaphrodite, actinomorphic; hypogynous, epigynous or perigynous; all floral parts usually freely inserted on the receptacle and cylindrically arranged.

Calyx of 5 free sepals, occasionally with an epicalyx (*Fragaria* spp.). Corolla usually of 5 free petals, alternating with the sepals. Sometimes petals are absent.

Androecium of variable number of stamens; usually 2, 3 or 4 times as many as the sepals.

Gynaecium of variable number of carpels; sometimes 1, or many apocarpous carpels or 5 syncarpous carpels. Ovary usually superior, but sometimes inferior (*Malus* spp.). Styles free and as many as there are carpels.

Nectaries found on the receptacle between the stamens and the carpels.

Flowers are mostly insect-pollinated; often protandrous.

Fruits vary; drupe in *Prunus* spp., collection of drupes in *Rubus* spp., follicles in *Spiraea* spp., collection of achenes in *Geum* spp., and false fruits in *Malus* spp., and *Pyrus* spp. Seeds non-endospermic.

British genera include:
   Crataegus (hawthorn)
   Fragaria (wild strawberry)
   Geum (avens)
   Rosa (wild rose)
   Rubus (bramble)

Genera cultivated by people for food and of great economic importance include:
   Fragaria (strawberry)
   Malus (apple)
   Prunus (almond, apricot, cherry, plum and sloe)
   Pyrus (pear)
   Rubus (raspberry, blackberry)

Rosa, Cotoneaster, Geum and *Spiraea* are cultivated as garden flowers.

DIVISION SPERMATOPHYTA
CLASS ANGIOSPERMAE
SUB-CLASS DICOTYLEDONES
FAMILY ROSACEAE

Malus sp.
Apple

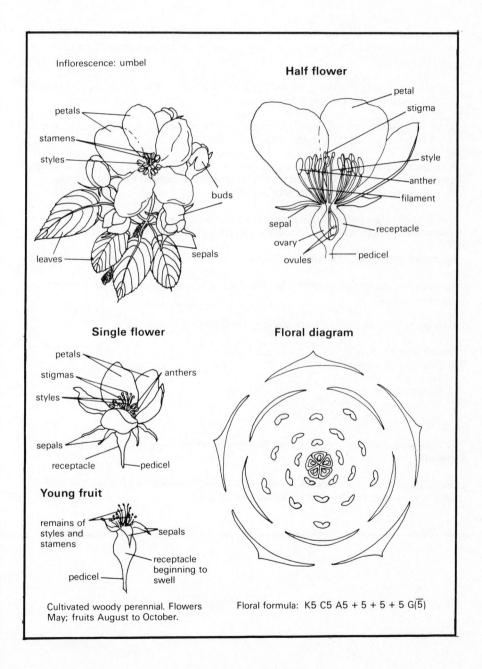

Inflorescence: umbel

petals
stamens
styles
buds
leaves
sepals

Half flower

petal
stigma
style
anther
filament
sepal
receptacle
ovary
ovules
pedicel

Single flower

petals
stigmas
styles
anthers
sepals
receptacle
pedicel

Young fruit

remains of
styles and
stamens
sepals
receptacle
beginning to
swell
pedicel

Floral diagram

Cultivated woody perennial. Flowers
May; fruits August to October.

Floral formula: K5 C5 A5 + 5 + 5 + 5 G($\overline{5}$)

Family: *Scrophulariaceae*

Mostly herbaceous, annual or perennial; with alternately or oppositely arranged often hairy leaves with no stipules. Some may be partial parasites (*Euphrasia* sp., eyebright).

Inflorescence racemose or cymose; flowers hermaphrodite, usually zygomorphic, hypogynous; floral parts cyclically arranged.

Calyx of 5, sometimes 4, sepals joined at the base. Corolla gamopetalous, of 5 petals; 5-lobed, or 2-lipped with indistinct lobes.

Androecium may be of 5 epipetalous stamens alternating with the corolla lobes; frequently 4 stamens (*Digitalis*) or fifth stamen represented by a staminode (*Scrophularia*); rarely 2 stamens (*Veronica*).

Gynaecium of 2 syncarpous carpels, bilocular with many ovules. Axile placentation. Ovary superior. Style simple with capitate or bi-lobed stigma.

Nectaries present at the base of the ovary.

Flowers are insect-pollinated by large insects such as bees and wasps; mostly protandrous.

Fruit a capsule, dehiscing by 2 valves; seeds endospermic.

British genera include:
   Digitalis (foxglove)
   Linaria (toadflax)
   Scrophularia (figwort)
   Verbascum (mullein)
   Veronica (speedwell)

Antirrhinum, *Calceolaria*, *Mimulus* and *Pentstemon* are all cultivated as garden flowers.

Some genera yield poisonous substances; *Digitalis* (foxglove) contains digitalin which is used in the treatment of heart conditions.

Digitalis purpurea
Foxglove

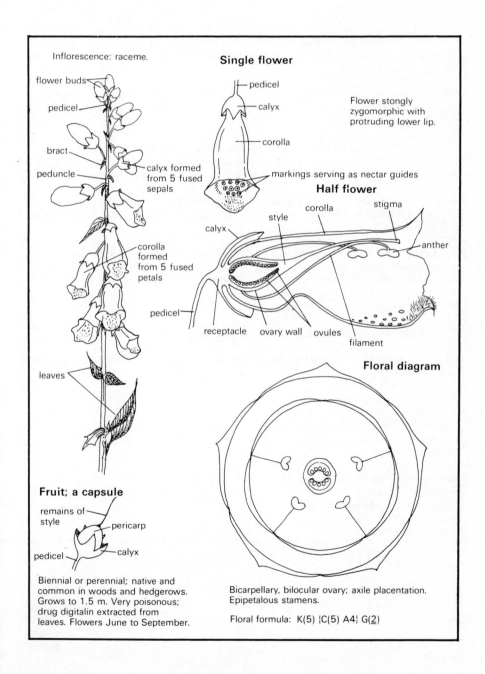

Inflorescence: raceme.

flower buds

pedicel

bract

peduncle

calyx formed from 5 fused sepals

corolla formed from 5 fused petals

leaves

Single flower

pedicel

calyx

corolla

markings serving as nectar guides

Flower stongly zygomorphic with protruding lower lip.

Half flower

corolla

stigma

style

calyx

anther

pedicel

receptacle   ovary wall   ovules

filament

Floral diagram

Fruit; a capsule

remains of style

pericarp

pedicel

calyx

Biennial or perennial; native and common in woods and hedgerows. Grows to 1.5 m. Very poisonous; drug digitalin extracted from leaves. Flowers June to September.

Bicarpellary, bilocular ovary; axile placentation. Epipetalous stamens.

Floral formula: K(5) ¦C(5) A4¦ G(2)

115

Family: Labiatae

Mostly herbaceous perennials and a few shrubs, possessing oppositely arranged often hairy and glandular leaves with no stipules; stems often hollow and square, swollen at the nodes.

Inflorescence cymose, contracted to form whorls often close together, forming spikes or heads; flowers usually hermaphrodite, zygomorphic, hypogynous; floral parts cyclically arranged.

Calyx gamosepalous, usually 5-toothed, may be 2-lipped with three upper and two lower sepals. Corolla gamopetalous, 4- or 5-lobed, distinctly 2-lipped; upper lip of 2 petals forming a hood, lower lip of 3 petals forming a platform.

Androecium usually of 4 epipetalous stamens alternating with the corolla segments; filaments all same length (*Mentha* spp., *Glechoma* spp.); sometimes only 2 stamens.

Gynaecium of 2 syncarpous carpels with 2 original loculi; later appearing quadrilocular due to growth of dividing wall; each loculus containing 1 ovule. Axile placentation. Ovary superior. Style simple with bi-lobed stigma.

Nectaries present at base of ovary.

Flowers are pollinated by insects with long tongues; mostly protandrous.

Fruit composed of 4 nutlets formed by separation of the 2 original loculi; known as a carcerulus.

British genera include:
   Glechoma (ground ivy)
   Lamium (deadnettle)
   Nepeta (catmint)
   Prunella (selfheal)

Some genera are cultivated by people as pot herbs and include:
   Mentha (mint)
   Origanum (marjoram)
   Salvia (sage)
   Thymus (thyme)

Lamium alba
White
deadnettle

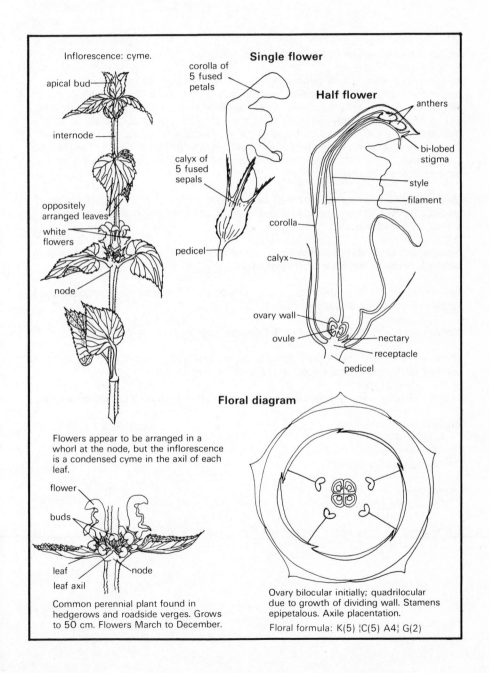

Inflorescence: cyme.

apical bud

internode

oppositely arranged leaves

white flowers

node

Single flower

corolla of 5 fused petals

calyx of 5 fused sepals

pedicel

Half flower

anthers

bi-lobed stigma

style

filament

corolla

calyx

ovary wall

ovule

nectary

receptacle

pedicel

Floral diagram

Flowers appear to be arranged in a whorl at the node, but the inflorescence is a condensed cyme in the axil of each leaf.

flower

buds

leaf

node

leaf axil

Common perennial plant found in hedgerows and roadside verges. Grows to 50 cm. Flowers March to December.

Ovary bilocular initially; quadrilocular due to growth of dividing wall. Stamens epipetalous. Axile placentation.

Floral formula: K(5) ¦C(5) A4¦ G(2)

Family: Compositae

Mostly herbaceous, annual or perennial, with alternately arranged, often hairy, leaves without stipules; often possessing latex.

Inflorescence a capitulum surrounded by an involucre of bracts; flowers small, referred to as florets, hermaphrodite or unisexual, usually actinomorphic, epigynous. Florets may be of two types: ligulate or tubular. Capitula may consist of one type of floret only, as in *Arctium* spp., *Cirsium* spp. and *Centaurea* spp. (tubular florets only), or as in *Hieracium* spp., *Lactuca* spp., and *Taraxacum* spp. (ligulate florets only). In capitula with both types of floret, the tubular florets are usually centrally situated (disc florets) surrounded by the ligulate florets (ray florets), e.g. *Bellis* spp.

Calyx usually absent or represented by a pappus of hairs, membranous scales, teeth or bristles. Corolla gamopetalous, 5-toothed or with lateral strap-shaped extension called a ligule.

Androecium of 5 epipetalous introrse stamens, the anthers usually fused to form a cylinder around the style (syngenesious).

Gynaecium of 2 syncarpous carpels, unilocular with 1 ovule. Ovary inferior. Style single with bi-lobed stigma.

Nectaries present at lower end of the corolla tube at the base of the style.

Flowers are mostly insect-pollinated; protandrous. Some genera produce seed without fertilization (apomixis), e.g. *Taraxacum*, *Hieracium*.

Fruit an achene, with pericarp and testa fused, often bearing a crown of pappus.

British genera include:
    Bellis (daisy)
    Arctium (burdock)
    Centaurea (cornflower)
    Cirsium (thistle)
    Hieracium (hawkweed)
    Taraxacum (dandelion)

Some genera are cultivated as garden flowers and include:
    Chrysanthemum, *Dahlia* and *Helianthus* (sunflower)

DIVISION SPERMATOPHYTA
CLASS ANGIOSPERMAE
SUB-CLASS DICOTYLEDONES
FAMILY COMPOSITAE

Taraxacum officinale
Dandelion

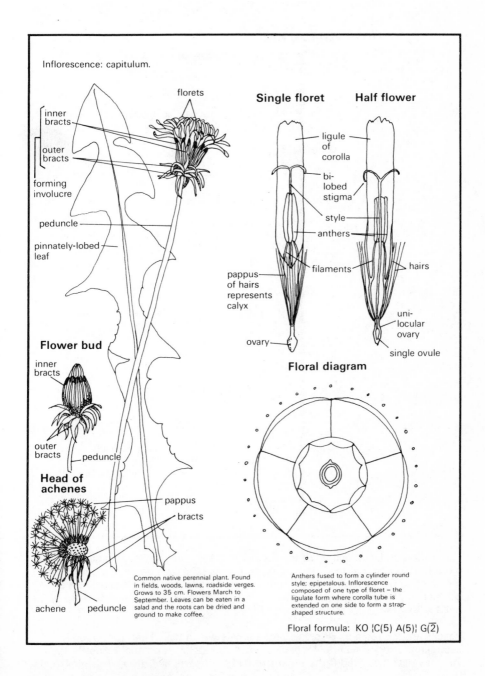

Inflorescence: capitulum.

florets

inner bracts

outer bracts

forming involucre

peduncle

pinnately-lobed leaf

Single floret   **Half flower**

ligule of corolla

bi-lobed stigma

style

anthers

filaments

hairs

pappus of hairs represents calyx

uni-locular ovary

ovary

single ovule

Flower bud

inner bracts

outer bracts

peduncle

Head of achenes

pappus

bracts

achene   peduncle

Floral diagram

Common native perennial plant. Found in fields, woods, lawns, roadside verges. Grows to 35 cm. Flowers March to September. Leaves can be eaten in a salad and the roots can be dried and ground to make coffee.

Anthers fused to form a cylinder round style; epipetalous. Inflorescence composed of one type of floret – the ligulate form where corolla tube is extended on one side to form a strap-shaped structure.

Floral formula: KO {C(5) A(5)} G($\overline{2}$)

The geological time scale

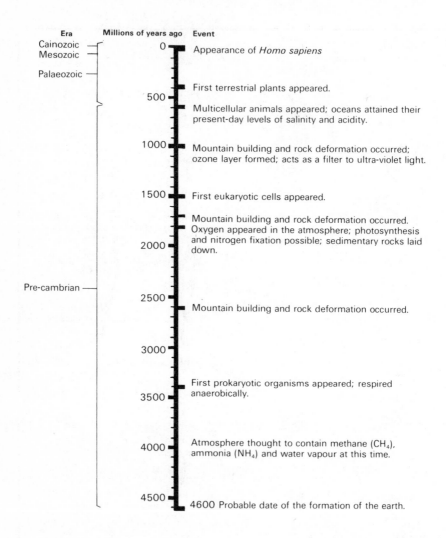

Era	Millions of years ago	Event
Cainozoic	0	Appearance of *Homo sapiens*
Mesozoic		
Palaeozoic		
		First terrestrial plants appeared.
	500	
		Multicellular animals appeared; oceans attained their present-day levels of salinity and acidity.
	1000	Mountain building and rock deformation occurred; ozone layer formed; acts as a filter to ultra-violet light.
	1500	First eukaryotic cells appeared.
		Mountain building and rock deformation occurred. Oxygen appeared in the atmosphere; photosynthesis and nitrogen fixation possible; sedimentary rocks laid down.
	2000	
Pre-cambrian	2500	Mountain building and rock deformation occurred.
	3000	
	3500	First prokaryotic organisms appeared; respired anaerobically.
	4000	Atmosphere thought to contain methane (CH_4), ammonia (NH_4) and water vapour at this time.
	4500	
		4600 Probable date of the formation of the earth.

The most significant events in the history of our planet are shown above, and a more detailed plan of the last 570 million years is shown on page 122. Most of the clues to the past life on earth come from fossils, and the classification and correlation of these fossils, together with the dating of rocks, have enabled us to build a hypothesis for the evolution of life.

Fossils are the remains of plants and animals which have been preserved in sedimentary rocks such as limestone, chalk, sandstone and shale. Fossilization will only have occurred if the organism had hard parts and decay was prevented;

it would then have to have been buried very quickly by sediment. This would have been a rare occurrence, and the chance that the fossil will be found is even rarer as not many of the fossil-bearing rocks are accessible to us.

The geological time scale
The last 570 million years

Period	Millions of years ago	Events
Quarternary	2	
Tertiary		Flowering plants dominant.
	65	
Cretaceous 100		Flowering plants appear in the fossil record.
	136	
Jurassic		Conifers dominant.
	190	
Triassic 200		Cycads appear; conifers flourish.
	225	
Permian		Conifers appear in the fossil record.
	280	
Carboniferous 300		Giant locopods and horsetails abundant, together with tree ferns; formed large forests in swampy regions.
	345	
Devonian		Land plants abundant; bryophytes, ferns, lycopods and horsetails all represented as fossils in these rocks; also fossils of fungi.
	395	
Silurian 400		First land plants appear in the fossil record; seaweeds abundant.
	430	
Ordovician		Photosynthetic bacteria and algae.
	500 500	
Cambrian		Photosynthetic bacteria and algae.
	570	
Pre-cambrian		Blue-green algae and bacteria.

actinomorphic – of angiosperm flowers, regular; capable of being cut vertically into two identical halves; radially symmetrical

adaxial – next to or towards the axis

adventitious – describes organs which appear in uncharacteristic positions e.g. roots developed on stems

alternation of generations – life cycle in which a diploid sporophyte generation producing haploid asexual spores by meiosis, alternates with a haploid gametophyte generation producing gametes. After fusion of the haploid gametes, the diploid sporophyte is restored

anastomosing – forming cross-connections; forming a network

androecium – in angiosperm flowers, the male reproductive organs, the stamens

anemophilous – in Spermatophyta, pollinated by wind or air movements

annual – in angiosperms, a plant which completes its life cycle in one growing season and then dies; yearly

annulus – in moss capsules and fern sporangia, a row of cells with thickening on all the walls except the outermost ones; involved in spore dispersal mechanism

anther – in angiosperm flowers, the portion of the stamen which bears the microsporangia or pollen sacs

antheridiophore – in Division Bryophyta, Class Hepaticae, Order Marchantiales, an upright, umbrella-shaped structure which develops on the male gametophyte plant and bears the antheridia on the upper surface of its disc

antheridium – a male sex organ or gametangium producing male gametes, or antherozoids

antherozoid – a motile male gamete characteristic of the algae, fungi, Bryophyta and Pteridophyta

apical – of, or at the apex, or tip, of a structure

apocarpous – in angiosperm flowers, a gynaecium made up of free carpels; the carpels not joined

apomixis – in Spermatophyta, reproduction by seeds which develop from unfertilized ovules

apophysis – in Division Bryophyta, Class Musci, the sterile tissue at the base of the spore capsule

archegoniophore – in Division Bryophyta, Class Hepaticae, Order Marchantiales, an upright structure which develops on the female gametophyte plant and bears archegonia on the lower surface of its disc

archegonium – a multicellular female sex organ or gametangium containing one or more oospheres

archesporial cell – a cell which gives rise to the spore mother cells in a sporangium

areola (plural: areolae) – in Division Bryophyta, Class Hepaticae, Order Marchantiales, the area surrounding the pores on the thallus upper surface

asexual – a form of reproduction in which new individuals are derived from a single parent, without fusion of gametes or nuclei; often involving the production of buds, spores or gemmae

autotrophic – describes organisms which synthesize their own organic requirements from simple inorganic compounds, using light or chemical reactions as sources of energy

Glossary

axil – the upper angle which a leaf makes with a stem; axillary buds may develop in this region

axile placentation – in angiosperm flowers, the margins of the carpels fold inwards, fusing together in the centre of the ovary to form a single central placenta

bark – refers to the tissues external to the wood of stems and roots of higher plants. The term may be used more specifically to refer to the protective tissues resulting from the activity of the cork cambium

basal – situated at the base or the lowest level

biennial – a plant which requires two growing seasons for the completion of its life cycle. Vegetative growth takes place during the first growing season, followed by flowering, seed production and death in the second growing season

biflagellate – having two flagella, as in the male gametes of Division Bryophyta and some members of the Division Pteridophyta

bifoliar spur – the two-needled dwarf shoot characteristic of the genus *Pinus*, Order Coniferales, Class Gymnospermae

bi-lobed – having two lobes

bract – in Division Bryophyta, Class Hepaticae, reduced leaf-like structures bearing antheridia in their axils; in angiosperms, a reduced leaf with a flower, or a branch of an inflorescence, in its axil

bracteole – a small bract

bract scale – in Division Spermatophyta, Class Gymnospermae, a scale in the female cone or strobilus: ovuliferous scales arise in the axils of bract scales

bulb – in angiosperms, an underground storage and perennating organ, which consists of swollen, fleshy leaf bases surrounding a short conical stem

bulbil – a small bulb-like structure of vegetative propagation

calyptra – in Division Bryophyta, the protective covering of the capsule, derived from the base of the archegonium

calyx – in angiosperms, the outermost part or whorl of the flower, composed of the sepals

cambium – a meristem; a layer of actively dividing cells, e.g. vascular cambium situated between the xylem and the phloem

capitate – having a head or knob

capsule – in Division Bryophyta, the spore-containing part of the sporogonium, or sporophyte generation

carinal canal – in Division Pteridophyta, Class Sphenopsida, a canal formed by the breakdown of the protoxylem

carotene – especially beta-carotene, an orange pigment found in all photosynthetic plants

carpel – in angiosperms, the structure in which the ovules are borne; the megasporophyll

chalaza – in Division Spermatophyta, the region of the ovule opposite the micropyle, where the nucellus, integuments and funicle meet

chlorophylls – green pigments common to all photosynthesizing plants; usually located in chloroplasts

chloroplast – a plastid containing photosynthetic pigments, especially chlorophylls, so that it looks green

Glossary

cilium (plural: cilia) – in Division Bryophyta, Class Musci, a segment of the peristome teeth

circinnate vernation – in ferns, the way in which the young fronds unroll from base to apex

coleoptile – in angiosperms, especially Monocotyledones, the protective sheath around the plumule of the embryo

coleorhiza – in angiosperms, especially Monocotyledones, the protective sheath around the radicle of the embryo

columella – in Division Bryophyta, Class Musci, the central sterile structure within the capsule

companion cell – in angiosperms, a small elongated cell with dense cytoplasm and a nucleus, found associated with sieve tubes in phloem tissue

cone – in some members of Division Pteridophyta and Division Spermatophyta, a shoot bearing sporophylls, usually packed closely together; a strobilus

convoluted – coiled, rolled together or twisted

corm – in angiosperms, an underground storage and perennating organ which consists of a short, upright swollen stem

corolla – in angiosperm flowers, the perianth whorl consisting of the petals of the flower

corona – in some monocotyledonous angiosperm flowers, a trumpet-shaped structure made up of the ligules of the perianth segments; especially *Pseudonarcissus* spp.

cortex – in stems and roots of vascular plants, the tissue between the epidermis and the vascular tissue

cotyledon – in Division Spermatophyta, the embryonic or first leaf, very different in form from subsequent leaves. The number of cotyledons varies according to the group, e.g. Monocotyledones have one, Dicotyledones have two

cupule – in angiosperms, a cup-shaped structure formed by bracts at the base of the fruit in certain genera

cuticle – waxy coating of cutin present on the epidermis of land plants; secreted by the epidermal cells

cutin – a waxy material composed of derivatives of fatty acids; component of the cuticle

cutinized – impregnated with cutin

cyme – in angiosperms, the type of branching or inflorescence derived from the growth of lateral buds, following abortion of the apical bud; the oldest flowers will occur at the top

dehiscence – the opening of sporangia or fruits by splitting

dichotomy – division or branching of an organ or structure into two equal portions; hence dichotomous

dictyostele – a stele composed of a network of strands; each strand consisting of a central portion of xylem surrounded by phloem; especially in Division Pteridophyta

dioecious – bearing male and female sex organs on separate plants

diploid – any nucleus, cell or organism which has twice the haploid number of chromosomes; often shown as 2n. All zygotes are diploid, as are the sporophytes in any alternation of generations.

disc – a flat, circular structure; the inner part of the capitulum in members of

Glossary

Family Compositae, Class Angiospermae

discoid – in the form of a disc; often descriptive of the shape of chloroplasts

dormant – in a resting condition; not growing; alive but with a very low metabolic rate; especially of seeds in Division Spermatophyta

dorsiventral – having distinct upper (dorsal) and lower (ventral) surfaces

dwarf shoot – in some Coniferales, especially *Pinus*, the short shoot on which the leaves are borne

elaters – elongated, hygroscopic structures with spiral bands of thickening which are formed in some sporangia and assist in spore dispersal; especially Division Bryophyta, Class Hepaticae

embryo – the young sporophyte derived from the diploid zygote

embryo sac – the mature female gametophyte in the angiosperms; contains eight nuclei, one of which is the egg cell

endocarp – in angiosperms, the inner layer of the pericarp, or fruit wall

endodermis – the inner layer of cells of the cortex of a stem or root

endosperm – in angiosperms, the nutritive tissue in the seed; sometimes used to describe the female prothallus tissue in the seeds of gymnosperms

entomophilous – in angiosperm flowers, insect-pollinated; having brightly-coloured flowers, scent, nectar to attract insects to the plant

epicalyx – in angiosperm flowers, small, sepal-like structures occurring in addition to the sepals, usually alternating with them; especially found in *Fragaria spp.* (strawberry)

epicarp – in angiosperms, the outer layer of the pericarp, or fruit wall; often forming a tough skin

epidermis – the outer layer of cells in plants

epigynous – in angiosperm flowers, where the perianth and androecium are inserted above the gynaecium; having an inferior ovary

epipetalous – in angiosperm flowers, especially monocotyledonous, descriptive of stamens with their filaments attached to the perianth segments

epiphyte – a plant which grows attached to another plant, but is not parasitic on it

exine – the outer layer of the wall of a pollen grain, or microspore

extrorse – in angiosperm flowers, with anthers facing outwards, away from the centre of the flower

filament – a chain of cells; cells arranged end to end; in angiosperm flowers, the stalk of the stamen

floret – a small flower; a single flower in a closely-packed inflorescence, e.g. Family Compositae

foliose – leafy; leaf-like

frond – a large leaf, usually pinnately divided; especially used to describe fern leaves

fruit – the ripened ovary of an angiosperm flower

funicle – in angiosperms, the stalk which attaches the ovule to the placenta in the ovary

gametangiophore – in Division Bryophyta, Class Hepaticae, a stalk which bears the gametangia in some thalloid liverworts; see archegoniophore, antheridiophore

gamete – a haploid sex cell which fuses with another haploid gamete to form a zygote

gametophyte – the haploid generation, which produces gametes

gamopetalous – in angiosperm flowers, having joined or united petals; petals fused together to form a corolla tube

gamosepalous – in angiosperm flowers, having joined or united sepals

gemma (plural: gemmae) – in Division Bryophyta, a small group of cells which can become detached from the parent plant and capable of developing into a new gametophyte; an organ of vegetative propagation

globose – spherical, like a globe

gynaecium – in angiosperms, the collective name for the female parts of the flower, the carpels

haploid – any nucleus, cell or organism having a single set of chromosomes; often shown as n

herbaceous – in Division Spermatophyta, having the characteristics of a herb; not undergoing much development of secondary vascular tissues

hermaphrodite – in angiosperms, having both stamens and carpels in the same flower; with both male and female reproductive organs on the same individual

heteromorphic – having more than one growth form; in an alternation of generations, the sporophyte will differ morphologically from the gametophyte

heterosporous – having spores of two sizes, microspores and megaspores

homologous – having a similar origin

homosporous – having spores of one size only

hyaline – clear, transparent

hygroscopic – capable of absorbing water

hypha – one of the filaments making up the vegetative body of a fungus; a single thread in a mycelium

hypocotyl – that part of an embryo or seedling between the radicle, or young root, and the cotyledon stalks

hypodermis – one or more layers of cells immediately below the epidermis; may be lignified as in *Pinus*

hypogynous – in angiosperm flowers, where the perianth and androecium are inserted below the gynaecium; having a superior ovary

indusium – in ferns, the protective covering of a sorus, consisting of a membranous flap of tissue

inferior – in angiosperm flowers, the gynaecium inserted below the other flower parts; the gynaecium of an epigynous flower

inflorescence – in angiosperms, a group of flowers borne on the same stalk or peduncle

integument – in Division Spermatophyta, the protective layer which develops around the ovule: in gymnosperms, usually only one develops; in angiosperms, two

intercalary – of meristems, situated between permanent tissues, e.g. in grasses, at the bases of the nodes

internode – that portion of the stem or axis between two nodes

intine – the inner layer of the microspore wall

introrse – in angiosperm flowers, with anthers facing inwards towards the centre of the flower

invagination – an intucking or infolding

Glossary

involucre – in Division Bryophyta, Class Hepaticae, especially thalloid genera, the protective flap which covers the developing archegonia; in foliose genera, the group of leaves surrounding the sex organs; in angiosperms, a group of bracts surrounding the inflorescence especially in Compositae

isobilateral – characteristic of monocotyledonous leaves; having both sides the same

isodiametric – of cells, having equal dimensions all round; of equal height, width and length

lamella – thin layer, membrane or plate-like structure

lamina – in higher plants, the leaf blade; the flattened part of the leaf attached to the petiole

lateral – borne or occurring at the side; as opposed to apical, at the apex

latex – a fluid produced by some flowering plants; a milky juice, especially in *Taraxacum* (dandelion)

leaf gaps – characteristic of ferns and seed plants, regions of the vascular cylinder of a stem immediately above the point of departure of a leaf trace; parenchyma differentiates instead of vascular tissue in these regions

leaf trace – a vascular bundle to a leaf

lignified – impregnated with lignin, a complex organic substance which provides resistance to compression

lignin – the major constituent of the cell walls of woody tissue; a polymer of sugars, amino acids and alcohols

ligule – in Division Pteridophyta, Class Lycopsida especially Selaginellales, a scale-like structure on the upper surface of the leaf; in grasses, an outgrowth at the junction of the leaf blade and the leaf sheath

loculus – in angiosperm flowers, a cavity in the ovary which contains ovules

macrophyllous – in ferns, having large leaves or fronds

malic acid – in ferns, an organic acid which attracts the antherozoids to the oosphere of an archegonium

marginal – in or on the margin or edge

marginal placentation – in angiosperm flowers, the carpels are fused at their margins and placentas occur as internal ridges on the ovary wall: alternatively known as parietal placentation

medulla – the central part of an organ

megasporangium – a sporangium in which megaspores develop

megaspore – a spore which gives rise to a female gametophyte

megasporophyll – a leaf-like structure bearing megasporangia

meiosis – reduction division of a nucleus in which the diploid (2n) number of chromosomes is halved, resulting in haploid (n) daughter nuclei

membranous – thin; translucent; papery

meristem – a group of undifferentiated cells retaining the ability to divide, resulting in growth and the formation of new tissues

mesocarp – in angiosperms, the middle layer of the pericarp, or fruit wall

microphyllous – in Division Pteridophyta, Class Lycopsida and Class Sphenopsida, having small leaves

micropylar – in seed plants, that end of the ovule nearest the micropyle

micropyle – in seed plants, the tiny pore between the integuments of the ovule; the usual site of entry of the pollen tube into the ovule

microsporangium – a sporangium in which microspores develop

microspore – a spore which gives rise to a male gametophyte

microsporophyll – a leaf-like structure bearing microsporangia

midrib – in Division Bryophyta, Class Hepaticae, the central, thicker region of the thallus, with rhizoids attached to the ventral surface; in leaves, the region of the main vein

mitosis – division of a nucleus so that the daughter nuclei have exactly the same number of chromosomes as the parent nucleus; can occur in both haploid and diploid cells

monoecious – with both male and female sex organs on the same plant

monopodial – descriptive of branching in which the main axis continues to grow indefinitely; the lateral branches are produced from axillary buds

mucilage – a viscous, complex carbohydrate substance produced by plants

multiflagellate – bearing many flagella

mycorrhiza – a symbiotic association between a fungus and the roots of a vascular plant

neck – the tube-like part of an archegonium

nectary – in angiosperm flowers, a gland which secretes the sugary fluid, nectar

nerve – in mosses, elongated cells in the centre of the leaf forming a midrib

node – the position on a stem or axis at which leaves or branches arise

nucleus – the organelle, present in eukaryotic cells, which contains the chromosomes

nucellus – in Division Spermatophyta, the central tissue of the ovule; in angiosperms, contains the embryo sac and is surrounded by the integuments

oogamy – the sexual fusion of two dissimilar gametes; the male gamete being small and unusually motile with very little stored food, and the female gamete being large and non-motile with larger food reserves

oosphere – a spherical, haploid female gamete; an ovum or egg cell; usually non-motile

oospore – a fertilized oosphere

operculum – in mosses, the lid of the capsule

ovary – in angiosperm flowers, the gynaecium consisting of one or more carpels, and containing the ovules

ovoid – egg-shaped

ovule – in seed plants, the structure that develops into the seed after fertilization

ovuliferous scale – in gymnosperms, the scale in the cone on the surface of which the ovules develop

palisade mesophyll – in leaves, the tissue consisting of regular columnar cells containing many chloroplasts, usually situated just below the upper epidermis

palm – a tree or shrub of the Palmae, a large group of monocotyledonous plants; having a crown of pinnate or fan-shaped leaves

palmate – having finger-like lobes arising from one point

pappus – in angiosperm flowers, especially Compositae, a ring of fine hairs developed from the calyx

paraphysis (plural: paraphyses) – a sterile hair consisting of a single row of cells; associated with the sex organs of the Division Bryophyta, Class Musci

parasite – an organism that lives on or in another living organism, called the host, from which it derives its food: the host organism does not benefit in any

Glossary

way and may be harmed by the presence of the parasite

parenchyma – a tissue consisting of unspecialized cells with thin cellulose cell walls and living contents

parenchymatous – composed of parenchyma tissue; like parenchyma

parietal placentation – see marginal placentation

parthenogenesis – a type of reproduction in which a female gamete can develop into a new individual without being fertilized by a male gamete

pedicel – in angiosperms, the stalk of an individual flower of an inflorescence

peduncle – in angiosperms, the flower stalk; the stalk of the inflorescence

pendulous – hanging loosely; dropping; suspended from the top

perennial – a plant that continues to grow from year to year

perianth – in Division Bryophyta, especially Jungermanniales, a tube-like structure surrounding an archegonium; in angiosperm flowers, the structures which enclose the reproductive organs: in Dicotyledones, there are usually two distinct whorls, the sepals and the petals; in Monocotyledones, both whorls are alike

pericarp – in angiosperms, the fruit wall, derived from the ovary wall

perichaetium – in Division Bryophyta, the tissue surrounding the archegonia in the Marchantiales: the leaves surrounding the sex organs in members of the Class Musci

pericycle – in higher plants, the tissue which lies between the endodermis and the vascular tissue

perigonial – in mosses, descriptive of leaf-like structures which surround the antheridia

perigynous – in angiosperm flowers, where the gynaecium is centrally situated on a concave receptacle, surrounded by the perianth and androecium on the rim; the ovary is superior

peristome – in mosses, the tooth-like structures which surround the opening of the capsule

petaloid – resembling a petal; petal-like

petals – in angiosperm flowers, the inner perianth segments, forming the corolla; usually brightly coloured and distinguishable from the green sepals

petiole – a leaf stalk, attaching the leaf to the stem

phenols – aromatic organic compounds found in the resin of gymnosperms

phloem – that part of the vascular tissue of higher plants, composed of living cells, through which organic substances are transported

pinna – a leaflet; the primary division of a pinnate leaf

pinnate – of compound leaves, having leaflets on either side of a common midrib

pinnule – a lobe of a leaflet of a pinnate leaf; subdivision of a pinna

pith – the central part of an organ; the medulla

placenta – in angiosperm flowers, the region of attachment of the ovules to the ovary wall

placentation – in angiosperms, the arrangement of the placentae on which the ovules are borne in the ovary

plantlet – a little plant

plumule – in seed plants, the shoot of the embryo

podium – in Division Pteridophyta, Class Lycopsida, the erect aerial branch bearing cones

pollen sac – in angiosperm flowers, a microsporangium, in which the pollen grains develop, borne on an anther

pollen tube – in seed plants, the tube formed when the pollen grain germinates; carries the male gametes to the egg cell

pollination – in seed plants, the transfer of pollen from the anther to the stigma of a flower

polyploid – having three or more sets of chromosomes per nucleus, e.g. 3n

primordium – a group of cells which gives rise to a tissue or an organ

prostrate – growing flat on the substratum; typical growth habit of some liverworts and mosses

protandrous – in angiosperm flowers, where the anthers mature before the carpels

prothallus – in Division Pteridophyta, the free-living gametophyte generation

protogynous – in angiosperm flowers, where the carpels mature before the anthers

protonema (plural: protonemata) – in mosses, a branched, multicellular filamentous structure resulting from the germination of a spore; new gametophyte plants develop from buds arising on the protonema

raceme – the type of branching or inflorescence in which flowers are borne on branches of the main axis; the older flowers occur at the base

rachis – the axis to which the leaflets are attached in a pinnate leaf

radicle – in seed plants, the root of the embryo

ramenta – in ferns, brown scaly epidermal hairs covering young leaves and stems

receptacle – in angiosperm flowers, the enlarged apex of the flower stalk from which the floral parts arise

resin – an acidic substance produced by a tree or shrub, e.g. *Pinus*

reticulate – like a network; applied to the venation typical of dicotyledonous leaves

rhizoid – a uni- or multicellular outgrowth which anchors a plant lacking roots to its substratum; especially gametophytes in Division Bryophyta and Division Pteridophyta

rhizome – an underground stem

rhizophore – in Division Pteridophyta, Class Lycopsida especially Selaginellales, an organ which bears the roots; intermediate in structure between a root and a stem

root stock – a short, vertical underground stem, bearing roots

saprophyte – an organism which obtains its food from dead or decaying organic matter

sclerenchyma – a mechanical tissue composed of thick-walled cells or fibres

secondary thickening – in angiosperms and gymnosperms, the formation of additional vascular tissue by the activity of lateral meristems; increase in diameter of stems and roots

seed – in Division Spermatophyta, the product of a fertilized ovule, consisting of an embryo and a food store enclosed by a protective seed coat, or testa

sepals – in angiosperm flowers, the outermost perianth segments, forming the calyx; usually distinguished from the petals by their green colour

Glossary

septum – a dividing wall or partition in a plant structure

sessile – having no stalk

seta – in Division Bryophyta, the stalk of the sporophyte capsule or sporogonium

sexual – a form of reproduction in which there is fusion of two haploid cells (gametes), or two haploid nuclei, to produce a diploid zygote

shrub – a woody plant with no definite main axis or trunk, and with branches arising near ground level

sieve cells – cells typical of phloem tissue; possess protoplasts but no nuclei

sieve tube – in angiosperm phloem tissue, a column of sieve-tube elements with perforated adjacent transverse end walls called sieve plates

sorus – in ferns, a group of sporangia

spathe – in monocotyledonous angiosperm flowers, a bract enclosing the inflorescence

spongy mesophyll – in leaves, the tissue made up of loosely arranged, irregular cells with large air spaces between them

sporangiophore – an erect stalk which bears a sporangium

sporangium – an asexual reproductive structure in which spores develop

spore – a reproductive structure, usually haploid, consisting of one or more cells

sporogonium – in Division Bryophyta, the spore-producing structure; the sporophyte generation

sporophyll – a leaf-like structure bearing sporangia

sporophyte – the diploid generation in a life cycle, producing haploid spores after meiosis

stamen – in angiosperm flowers, the anther, bearing microsporangia, and the filament; the androecium of the flower

staminode – in angiosperm flowers, an infertile stamen, may be modified or reduced

starch – an insoluble polysaccharide, which forms the principal reserve food material in green plants; a product of photosynthesis

stele – the primary vascular tissue of a root or stem, consisting of xylem, phloem, pericycle and endodermis (if present)

stellate – star-shaped

stigma – in angiosperm flowers, the part of the gynaecium which is receptive to the pollen; the surface on which the pollen germinates; often borne at the apex of a style

stipulate – having stipules

stipule – in angiosperms, a leaf-like outgrowth from the base of a petiole; usually occur in pairs

stoma (plural: stomata) – in terrestrial plants, a pore in the epidermis surrounded by a pair of guard cells; a pore through which gaseous exchange occurs

stomium – in fern sporangia, the region of thin-walled cells at which dehiscence occurs

strobilus – see cone

style – in angiosperm flowers, a sterile portion of the gynaecium which connects the stigma with the ovary

substratum – the material or substance on which an organism grows

subterranean – underground

superior – in angiosperm flowers, the gynaecium inserted above the other floral parts; the gynaecium of a hypogynous flower

suspensor – a group of cells arising from a zygote, which pushes the young embryo into the nutritive tissue

sympodial – branching in which the terminal bud or the apex of the primary axis stops growing; further growth is achieved by the development of apical buds on lateral branches

syncarpous – in angiosperm flowers, the gynaecium consisting of fused carpels

tapetum – a layer of nutritive cells in the sporangium, surrounding the spore mother cells

tendril – in angiosperms, a modified stem, leaf or part of a leaf, by means of which climbing plants gain attachment to a support

terpenes – hydrocarbons derived from the resin of gymnosperms, e.g. turpentine

testa – in angiosperms, the seed coat derived from the integuments which surrounded the ovule

tetrad – a group of four cells produced by a meiotic division

thallus – a simple plant body showing no differentiation into stem, leaf and root; typical of some genera in Division Bryophyta, Class Hepaticae

thalloid – resembling a thallus; having a simple, flat shape

trabeculate – having bars or elongated cells crossing a cavity

tracheid – in xylem tissue, an elongated, lignified pitted cell with oblique end walls

tree – a woody plant with a definite main axis or trunk, and with branches well above ground level

trifoliate – with three leaves, or leaflets

tri-radiate – three-rayed; having three radiating lines or parts

tuber – a rounded storage organ formed from a root or a stem

tuberous – swollen; resembling a tuber

vacuolated – having vacuoles; possessing fluid-filling cavities

vallecular canals – in Division Pteridophyta, Class Sphenopsida, air-filled longitudinal canals in the cortex of the stem

valve – in liverworts, one of several sections into which the capsule splits when dehiscing

vascular tissue – in higher plants, the conducting tissue, made up of phloem and xylem

vein – a vascular strand in a leaf

venation – the arrangement of veins in a leaf

venter – in Division Bryophyta and Division Pteridophyta, the basal region of an archegonium, containing the oosphere

versatile – in angiosperm flowers, descriptive of anthers that swing freely, the anther being fixed by its back to the tip of the filament

vessel – in angiosperms, a tube in the xylem tissue, formed from a column of lignified cells, the vessel segments, whose end walls have broken down; water-transporting function

viable – capable of living; capable of germinating

whorl – a group of similar structures arising from the same level on a stem, forming a circle round it, e.g. petals and sepals in angiosperm flowers and leaves in horsetails

xanthophylls – a group of carotenoid pigments which absorb light in the blue-

Glossary

violet range; found in all photosynthetic plants

xylem – a composite tissue in higher plants, with a water-conducting function and also contributing to the mechanical support of the plant

zoidogamous – of fertilization which is dependent on free-swimming male gametes fusing with female gametes

zygomorphic – bilaterally symmetrical

zygote – a diploid cell which results from the fusion of two gametes